Handbook of Nuclear Power

Handbook of Nuclear Power

Alex Romerro

STATES
ACADEMIC PRESS
www.statesacademicpress.com

Published by States Academic Press,
109 South 5th Street,
Brooklyn, NY 11249, USA
www.statesacademicpress.com

Handbook of Nuclear Power
Alex Romerro

International Standard Book Number: 978-1-63989-264-8 (Hardback)

Cataloging-in-Publication Data

Handbook of nuclear power / Alex Romerro.
p. cm.
Includes bibliographical references and index.
ISBN 978-1-63989-264-8
1. Nuclear energy. 2. Nuclear engineering. 3. Nuclear power plants.
4. Nuclear facilities. I. Romerro, Alex.

TK9145 .H36 2022
621.48--dc23

TABLE OF CONTENTS

PREFACE

The use of nuclear energy released during nuclear reactions for the generations of electricity is termed as nuclear power. The heat released during nuclear reactions is used to operate steam turbines in a nuclear power plant. Nuclear reactions can be classified into nuclear fusion, nuclear decay and nuclear fission. Uranium and plutonium are the primarily used fuels in nuclear facilities. Nuclear fission of uranium is the most commonly employed technique for generation of heat in nuclear power plants. They are very much similar to a coal fired power plant except for the boiler which is replaced by a reactor. Nuclear power has one of the lowest fatality rates per unit of energy generated and causes significant reduction in the emission of greenhouse gases. This book is a compilation of chapters that discuss the most vital concepts in the field of nuclear power. Most of the topics introduced herein cover new techniques and the applications of nuclear power. As this field is emerging at a rapid pace, the contents of this book will help the readers understand the modern concepts and applications of the subject.

To facilitate a deeper understanding of the contents of this book a short introduction of every chapter is written below:

Chapter 1- Nuclear power is the use of nuclear reactions to generate heat which is usually used in steam turbines to produce electricity. It is both clean and efficient. Nuclear energy, nuclear fuel, nuclear power plants, nuclear reactor, etc. are a few topics which are related to nuclear power. This chapter serves as an introduction to nuclear power.

Chapter 2- Nuclear reactors that use water as a coolant and moderator are called water-cooled reactors. There are different types of water-cooled reactors. Boiling water reactor, pressurized water reactor, water-water energetic reactor, RBMK reactor, etc. are a few examples. This chapter deals with the various types of water-cooled reactors and examines them in a detailed manner for an extensive understanding of the subject.

Chapter 3- Gas-cooled reactors are those nuclear reactors which use graphite as a neutron moderator and carbon dioxide as a coolant. They offer potential operational and safety benefits over water-cooled reactors. Magnox reactor, UNGG reactor, advanced gas-cooled reactor, high temperature cooled reactor and gas-cooled fast reactor are a few gas-cooled reactors that are discussed thoroughly in this chapter.

Chapter 4- A breeder reactor is a nuclear reactor which produces more fissile materials than it consumes to generate energy. They are designed to extend the nuclear fuel supply for the generation of electricity. Fast breeder reactors use fast neutrons to generate nuclear fuels and power. In thermal breeder reactors the fission chain is sustained by thermal electrons. The topics elaborated in this chapter will help in gaining a better perspective about breeder reactors and their types.

Chapter 5- A nuclear reaction is a process in which two nuclei or nuclear particles collide to produce different products than the initial particles. Nuclear fusion and nuclear fission are two types of nuclear reactions. Fusion is the process where two nuclei combine together and fission is the splitting of a heavy nucleus into two lighter nuclei. Nuclear reactions and their various aspects like fusion, fission, elastic scattering, inelastic scattering, capture reaction, etc. are covered extensively in this chapter.

Chapter 6- Heat flow process deals with the generation, transfer, use and cooling of thermal energy within a system. Heat generation, neutron flux, steam generator, heat transport, emergency core cooling system, moderator and moderator system are some of the aspects that fall under its domain. This chapter has been carefully written to provide an easy understanding of all these topics with relation to their implementation in nuclear reactors.

Chapter 7- Nuclear safety is the achievement of proper operating conditions and prevention or mitigation of accident consequences, resulting in protection of workers, the public and the environment from undue radiation hazards. This chapter provides a detailed analysis into the critical topic of nuclear safety, its concepts and underlying principles and related techniques for preventive, protective and fail-safe measures.

I would like to share the credit of this book with my editorial team who worked tirelessly on this book. I owe the completion of this book to the never-ending support of my family, who supported me throughout the project.

Alex Romerro

Chapter 1

An Introduction to Nuclear Power

Nuclear power is the use of nuclear reactions to generate heat which is usually used in steam turbines to produce electricity. It is both clean and efficient. Nuclear energy, nuclear fuel, nuclear power plants, nuclear reactor, etc are a few topics which are related to nuclear power. This chapter serves as an introduction to nuclear power.

Nuclear power is a clean and efficient way of boiling water to make steam, which turns turbines to produce electricity. Nuclear power plants use low-enriched uranium fuel to produce electricity through a process called fission—the splitting of uranium atoms in a nuclear reactor. Uranium fuel consists of small, hard ceramic pellets that are packaged into long, vertical tubes. Bundles of this fuel are inserted into the reactor.

A single uranium pellet, slightly larger than a pencil eraser, contains the same energy as a ton of coal, 3 barrels of oil, or 17,000 cubic feet of natural gas. Each uranium fuel pellet provides up to five years of heat for power generation. And because uranium is one of the world's most abundant metals, it can provide fuel for the world's commercial nuclear plants for generations to come.

Nuclear power offers many benefits for the environment as well. Power plants don't burn any materials so they produce no combustion by-products. Additionally, because they don't produce greenhouse gases, nuclear plants help protect air quality and mitigate climate change.

When it comes to efficiency and reliability, no other electricity source can match nuclear. Nuclear power plants can continuously generate large-scale, around-the-clock electricity for many months at a time, without interruption.

Currently, nuclear energy supplies 12 percent of the world's electricity and approximately 20 percent of the energy in the United States. A total of 30 countries worldwide are operating 450 nuclear reactors for electricity generation.

Principles of Nuclear Power

Atoms are constructed like miniature solar systems. At the center of the atom is the nucleus; orbiting around it are electrons. The nucleus is composed of protons and neutrons, very densely packed together. Hydrogen, the lightest element, has one proton; the heaviest natural element, uranium, has 92 protons.

The nucleus of an atom is held together with great force, the "strongest force in nature." When bombarded with a neutron, it can be split apart, a process called fission (pictured to the right). Because uranium atoms are so large, the atomic force that binds it together is relatively weak, making uranium good for fission.

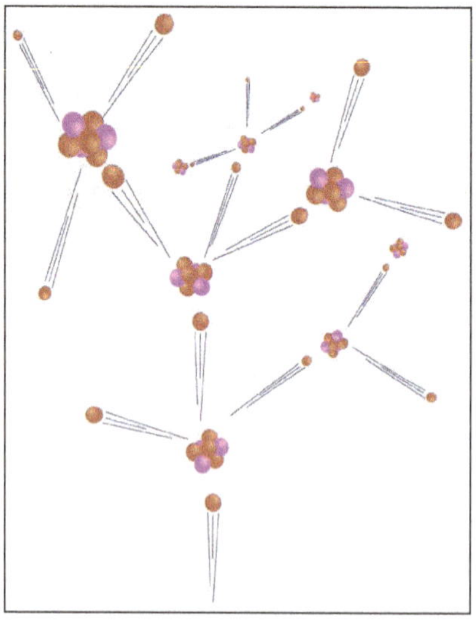

Figure: During fission, a neutron bombards a uranium atom, releasing more neutrons and triggering a chain reaction.

In nuclear power plants, neutrons collide with uranium atoms, splitting them. This split releases neutrons from the uranium that in turn collide with other atoms, causing a chain reaction. This chain reaction is controlled with "control rods" that absorb neutrons.

In the core of nuclear reactors, the fission of uranium atoms releases energy that heats water to about 520 degrees Farenheit. This hot water is then used to spin turbines that are connected to generators, producing electricity.

Figure: An open pit uranium mine in Namibia.

Mining and Processing Nuclear Fuels

Uranium is one of the least plentiful minerals—making up only two parts per million in the earth's crust—but because of its radioactivity it is a plentiful supply of energy. One pound of uranium has as much energy as three million pounds of coal.

Radioactive elements gradually decay, losing their radioactivity. The time it takes to lose half of its radioactivity is called a "half-life." U-238, the most common form of uranium, has a half-life of 4.5 billion years.

Uranium is found in a number of geological formations, as well as sea water. To be mined as a fuel, however, it must be sufficiently concentrated, making up at least one hundred parts per million (0.01 percent) of the rock it is in.

The mining process is similar to coal mining, with both open pit and underground mines. It produces similar environmental impacts, with the added hazard that uranium mine tailings are radioactive. Groundwater can be polluted not only from the heavy metals present in mine waste, but also from the traces of radioactive uranium still left in the waste. Half of the people employed by the uranium mining industry work on cleaning up the mines after use.

Uranium comes in two forms, U-235 and U-238. As found in nature, uranium is more than 99 percent U-238; unfortunately, U-235 is what is used in power plants. U-238 can also be processed into plutonium, which is also fissionable.

Once mined, the uranium ore is sent to a processing plant to be concentrated into a useful fuel. There are 16 processing plants in the US, although eight are inactive. Most uranium concentrate is made by leaching the uranium from the ore with acids. (Sometimes the concentrate is made underground, without removing the uranium ore.) When finished, the uranium ore is turned into U3O8, the fuel form of uranium, and shaped into small pellets.

The pellets are then packed into 12-foot long rods, called fuel rods. The rods are bundled together into fuel assemblies, ready to be used in the core of a reactor.

Nuclear Reactors

There are currently 99 commercial nuclear reactors in operation in the United States. Over a dozen commercial reactors have been shut down permanently, with more retirements likely to be announced in coming years.

Most of the plants in operation are "light water" reactors, meaning they use normal water in the core of the reactor. Different reactor technologies are in use abroad, such as the "heavy water" reactors in Canada. U.S. reactors account for more than one-fourth of nuclear power capacity in the world.

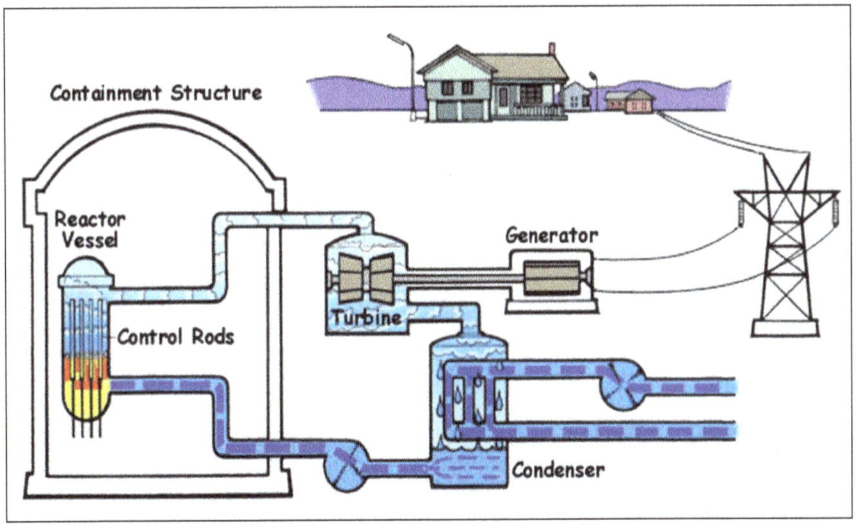

In the United States, two-thirds of the reactors are pressurized water reactors (PWR) and the rest are boiling water reactors (BWR). In a boiling water reactor, shown above, the water is allowed to boil into steam, and is then sent through a turbine to produce electricity.

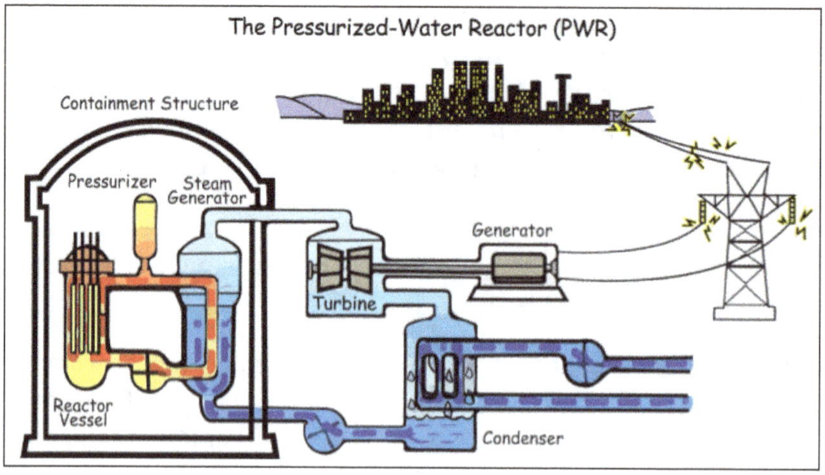

In pressurized water reactors, the core water is held under pressure and not allowed to boil. The heat is transferred to water outside the core with a heat exchanger (also called a steam generator), boiling the outside water, generating steam, and powering a turbine. In pressurized water reactors, the water that is boiled is separate from the fission process, and so does not become radioactive.

After the steam is used to power the turbine, it is cooled off to make it condense back into water. Some plants use water from rivers, lakes or the ocean to cool the steam, while others use tall cooling towers. The hourglass-shaped cooling towers are the familiar landmark of many nuclear plants. For every unit of electricity produced by a nuclear power plant, about two units of waste heat are rejected to the environment.

Commercial nuclear power plants range in size from about 60 megawatts for the first generation of plants in the early 1960s, to over 1000 megawatts. Many plants contain more than one reactor. The Palo Verde plant in Arizona, for example, is made up of three separate reactors, each with a capacity of 1,334 megawatts.

Some foreign reactor designs use coolants other than water to carry the heat of fission away from the core. Canadian reactors use water loaded with deuterium (called "heavy water"), while others are gas cooled. One plant in Colorado, now permanently shut down, used helium gas as a coolant (called a High Temperature Gas Cooled Reactor). A few plants use liquid metal or sodium.

Figure: Experimental tunnels at the Yucca Mountain waste repository site.

Nuclear Waste

Over 67,000 metric tons of highly radioactive waste had been produced by US-based nuclear reactors. That increases by about 2,000 metric tons every year. Before the mid-1970s, the plan for spent uranium was to reprocess it into new fuel.

Since a by-product of reprocessing is plutonium, which can be used to make nuclear weapons, Reprocessing also had a difficult time competing economically with new uranium fuel. Since then, the Department of Energy has been studying storage sites for long-term burial of the waste.

Meanwhile, radioactive waste is being stored at the nuclear plants where it is produced. The most common option is to store it in spent fuel cooling pools, large steel-lined tanks that use electricty to circulate water. As these pools fill up, some fuel rods are being transferred to large steel and concrete casks, which are considered safer.

In addition to the spent fuel, the plants themselves contain radioactive waste that must be disposed of after they are shut down. Plants can either be disassembled immediately or can be kept in storage for a number of years to give the radiation some time to diminish. Most of the plant is considered "low level waste" and can be stored in less secure locations.

Currently, only two sites accept low level waste: Barnwell in South Carolina and Hanford in Washington. NRC estimates decommissioning costs to range from $133 million to $303 million per plant, but so far no large reactors have been decommissioned. A number are in storage waiting decommissioning at a future time.

The Rise of Nuclear Power

The principles of nuclear power were formulated by physicists in the early 20th century. In 1939, German scientists discovered the process of fission, triggering a race with US scientists to use the incredible power of fission to create a bomb.

Through the intense effort of the Manhattan Project, the atomic bomb was created by 1945, and used to destroy Hiroshima and Nagasaki at the end of World War II. After the war, "great atomic power" was seen as a potential new energy source. The government's Plowshare Program thought atomic explosions would be a labor-saving way to dig canals and drinking water reservoirs and to mine for gas and oil.

Nuclear Energy

Nuclear energy is the energy in the nucleus, or core, of an atom. Atoms are tiny units that make up all matter in the universe, and energy is what holds the nucleus together. There is a huge amount of energy in an atom's dense nucleus. In fact, the power that holds the nucleus together is officially called the "strong force." Nuclear energy can be used to create electricity, but it must first be released from the atom. In the process of nuclear fission, atoms are split to release that energy.

A nuclear reactor, or power plant, is a series of machines that can control nuclear fission to produce electricity. The fuel that nuclear reactors use to produce nuclear fission is pellets of the element uranium. In a nuclear reactor, atoms of uranium are forced to break apart. As they split, the atoms release tiny particles called fission products. Fission products cause other uranium atoms to split, starting a chain reaction. The energy released from this chain reaction creates heat.

The heat created by nuclear fission warms the reactor's cooling agent. A cooling agent is usually water, but some nuclear reactors use liquid metal or molten salt. The cooling agent, heated by nuclear fission, produces steam. The steam turns turbines, or wheels turned by a flowing current. The turbines drive generators, or engines that create electricity.

Rods of material called nuclear poison can adjust how much electricity is produced. Nuclear poisons are materials, such as a type of the element xenon, that absorb some of the fission products created by nuclear fission. The more rods of nuclear poison that are present during the chain reaction, the slower and more controlled the reaction will be. Removing the rods will allow a stronger chain reaction and create more electricity.

About 15 percent of the world's electricity is generated by nuclear power plants. Nations such as Lithuania, France, and Slovakia create almost all of their electricity from nuclear power plants.

Nuclear Food Uranium

Uranium is the fuel most widely used to produce nuclear energy. That's because uranium atoms split apart relatively easily. Uranium is also a very common element, found in rocks all over the world. However, the specific type of uranium used to produce nuclear energy, called U-235, is rare. U-235 makes up less than one percent of the uranium in the world. Once uranium is mined, it must be extracted from other minerals. It must also be processed before it can be used.

Because nuclear fuel can be used to create nuclear weapons as well as nuclear reactors, only nations that are part of the Nuclear Non-Proliferation Treaty (NPT) are allowed to import uranium or plutonium, another nuclear fuel. The treaty promotes the peaceful use of nuclear fuel, as well as limiting the spread of nuclear weapons.

A typical nuclear reactor uses about 200 tons of uranium every year. Complex processes allow some uranium and plutonium to be re-enriched or recycled. This reduces the amount of mining, extracting, and processing that needs to be done.

Nuclear energy produces electricity that can be used to power homes, schools, businesses, and hospitals. The first nuclear reactor to produce electricity was located near Arco, Idaho. The Experimental Breeder Reactor began powering itself in 1951. The first nuclear power plant designed to provide energy to a community was established in Obninsk, Russia, in 1954.

Building nuclear reactors requires a high level of technology, and only the countries that have signed the Nuclear Non-Proliferation Treaty can get the uranium or plutonium that is required. For these reasons, most nuclear power plants are located in the developed world.

Nuclear power plants produce renewable, clean energy. They do not pollute the air or release greenhouse gases. They can be built in urban or rural areas, and do not radically alter the environment around them.

The steam powering the turbines and generators is ultimately recycled. It is cooled down in a separate structure called a cooling tower. The steam turns back into water and can be used again to produce more electricity. Excess steam is simply recycled into the atmosphere, where it does little harm as clean water vapor.

However, the byproduct of nuclear energy is radioactive material. Radioactive material is a collection of unstable atomic nuclei. These nuclei lose their energy and can affect many materials around them, including organisms and the environment. Radioactive material can be extremely toxic, causing burns and increasing the risk for cancers, blood diseases, and bone decay.

Radioactive waste is what is left over from the operation of a nuclear reactor. Radioactive waste is mostly protective clothing worn by workers, tools, and any other material that have been in contact with radioactive dust. Radioactive waste is long-lasting. Materials like clothes and tools can stay radioactive for thousands of years. The government regulates how these materials are disposed of so they don't contaminate anything else.

Used fuel and rods of nuclear poison are extremely radioactive. The used uranium pellets must be stored in special containers that look like large swimming pools. Water cools the fuel and insulates the outside from contact with the radioactivity. Some nuclear plants store their used fuel in dry storage tanks above ground.

The storage sites for radioactive waste have become very controversial in the United States. For years, the government planned to construct an enormous nuclear waste facility near Yucca Mountain, Nevada, for instance. Environmental groups and local citizens protested the plan. They worried about radioactive waste leaking into the water supply and the Yucca Mountain environment, about 130 kilometers (80 miles) from the large urban area of Las Vegas, Nevada. Although the government began investigating the site in 1978, it stopped planning for a nuclear waste facility in Yucca Mountain in 2009.

Critics of nuclear energy worry that the storage facilities for radioactive waste will leak, crack, or erode. Radioactive material could then contaminate the soil and groundwater near the facility. This could lead to serious health problems for the people and organisms in the area. All communities would have to be evacuated.

This is what happened in Chernobyl, Ukraine, in 1986. A steam explosion at one of the power plants four nuclear reactors caused a fire, called a plume. This plume was highly radioactive, creating a cloud of radioactive particles that fell to the ground, called fallout. The fallout spread over the Chernobyl facility, as well as the surrounding area. The fallout drifted with the wind, and the particles entered the water cycle as rain. Radioactivity traced to Chernobyl fell as rain over Scotland and Ireland. Most of the radioactive fallout fell in Belarus.

The environmental impact of the Chernobyl disaster was immediate. For kilometers around the facility, the pine forest dried up and died. The red color of the dead pines earned this area the nickname the Red Forest. Fish from the nearby Pripyat River had so many radioactivities that people could no longer eat them. Cattle and horses in the area died.

More than 100,000 people were relocated after the disaster, but the number of human victims of Chernobyl is difficult to determine. The effects of radiation poisoning only appear after many years. Cancers and other diseases can be very difficult to trace to a single source.

Nuclear reactors use fission, or the splitting of atoms, to produce energy. Nuclear energy can also be produced through fusion or joining (fusing) atoms together. The sun, for instance, is constantly undergoing nuclear fusion as hydrogen atoms fuse to form helium. Because all life on our planet depends on the sun, you could say that nuclear fusion makes life on Earth possible.

Nuclear power plants do not have the capability to safely and reliably produce energy from nuclear fusion. It's not clear whether the process will ever be an option for producing electricity. Nuclear engineers are researching nuclear fusion, however, because the process will likely be safe and cost-effective.

Advantages of Nuclear Energy

Low Pollution

Nuclear power also has a lot fewer greenhouse emissions. It has been determined that the number of greenhouse gases have decreased by almost half because of the prevalence in the utilization of nuclear power.

This avoids more than 470 million metric tons of carbon each year, which is the equivalent of removing 100 million cars off of the road. The thermal energy from nuclear reactors may also be used to decarbonize other energy-intensive sectors such as transportation, the largest contributor to carbon pollution.

Nuclear energy has the least effect on nature since it doesn't discharge any gasses like methane and carbon dioxide, which are the primary "greenhouse gasses." There is no unfavorable impact on water, land or any territories because of the utilization of nuclear power, except in times where transportation is utilized.

The average emissions for nuclear are 29 tonnes of CO_2 per gigawatt-hour (GWh) of energy produces. This compares favorably with renewable sources like solar (85 tonnes per GWh) and wind (26 tonnes per GWh) and even more favorably with fossil fuels like lignite (1,054 tonnes per GWh) and coal (888 tonnes per GWh).

High Power Output

The fuel to power output ratio for nuclear energy is incredibly high. It has the capacity to meet city and industrial needs with just one reactor, let alone multiple. A relatively small amount of uranium can be used to fuel a 1000 Megawatts electric plant, thus providing enough electricity to power a city of about half a million people.

Renewable sources, such as solar and wind, provide only enough power to meet residential or office needs. They don't yet have the capacity of nuclear to handle large-scale power needs, especially in the manufacturing world.

Stable Base Load Energy

Nuclear power plants provide a stable baseload of energy. Nuclear energy is widely used in America and makes up around 20% of all electricity generated in the United States. This efficient energy source comes from the 98 nuclear power reactors dotted around 30 different states in the US.

The stable production of power created by nuclear power plants means that it can ideally be used in conjunction with other forms of renewable energy. For example, when the wind is blowing, nuclear plants can adjust energy output to be lower. Conversely, when the wind is not blowing, and greater energy is needed, nuclear energy can be adjusted to compensate for the lack of wind (or solar) generated power.

Low Operating Costs

Nuclear power produces very inexpensive electricity and cheaper than gas, coal, or any other fossil fuel plants. The cost of the uranium, which is utilized as a fuel in this process, is low, and it is needed very little to produce massive power. Also, even though the expense of setting up nuclear power plants is moderately high, the expense of running them is quite low.

It has been estimated that even factoring in costs such as managing radioactive fuel and disposal nuclear plants cost between 33 to 50 percent of a coal plant and 20 to 25 percent of a gas combined-cycle plant.

The normal life of a nuclear reactor is anywhere from 40-60 years, depending on how often it is used and how it is being used. These variables, when consolidated, make the expense of delivering power low. Even if the cost of uranium goes up, the impact on the cost of power will be that much lower.

That to replace a 1GW nuclear power plant would require 2GW of coal or 3GW to 4GW from renewable sources to generate the same amount of electricity. Also, the impact on the cost of power will be that much lower.

Reliability

It is estimated that with the current rate of consumption of uranium, we have enough uranium for another 70-80 years. A nuclear power plant when in the mode of producing energy can run uninterrupted for even a year and more without interruptions or maintenance, making it a more reliable source of energy.

As solar and wind energy are dependent upon weather conditions, the nuclear power plant has no such constraints and can run without disruption in any climatic condition. The consistent criticism of renewable energies, e.g., wind and solar energy are that they only produce power when the wind is blowing, or the sun is shining.

There are sure monetary focal points in setting up nuclear power plants and utilizing nuclear energy in place of traditional energy. It is one of the significant sources of power all through the country.

The best part is that this energy has a persistent supply. It is broadly accessible, there is a lot in storage, and it is believed that the supply is going to last much, much longer than that of fossil fuels that are used in the same capacity.

More Proficient than Fossil Fuels

The other primary point of interest in utilizing nuclear energy is that it is more compelling and more proficient than other energy sources. A number of nuclear energy innovations have made it a much more feasible choice than others.

They have high energy density as compared to fossil fuels. The amount of fuel required by the nuclear power plant is comparatively less than what is required by other power plants as the energy released by nuclear fission is approximately ten million times greater than the amount of energy released by fossil fuel atom.

It doesn't Rely on Fossil Fuels

This is one of the reasons that numerous nations are putting a lot of time and money into nuclear power. So what's nuclear power's greatest benefit, above any other benefit that we may explore? It doesn't rely on fossil fuels and isn't influenced by fluctuating oil and gas costs. Coal and natural gas power plants discharge carbon dioxide into the air, which causes a number of environmental issues. With nuclear power plants, carbon emissions are insignificant.

Renewable

Nuclear energy is not a renewable resource. Uranium, the nuclear fuel that is used to produce nuclear energy, is limited and cannot be produced again and again on demand. However, uranium reserves are estimated to last another 80 years, whereas fossil fuels have a much more limited lifespan. Since the Industrial Revolution, humans have been consistently and constantly depleting our fossil fuel reserves. If we continue consuming fossil fuels and keep increasing our consumption as the world population grows, the world is estimated to run out of oil by 2052, gas in 2060, and coal by 2088.

On the other hand, by using breeder and fusion reactors, we can produce other fissionable elements. One such element is called plutonium that is produced by the by-products of chain-reaction. Also, if we know how to control atomic fusion, the same reactions that fuel the sun, we can have almost unlimited energy. Thorium is a greener alternative that has lately been come to notice. China, Russia and India already have plans to start using thorium to fuel their reactors in the near future.

Economic Impact

Nuclear power provides many benefits to the economy with the number of jobs and prosperity a new plant brings. A new nuclear plant creates 400 to 700 permanent jobs and also thousands of others during its construction. Most nuclear sites have at least 2 plants. Whereas jobs created elsewhere are just 90 jobs for a coal plant, and 50 for a natural gas plant.

Each facility generates close to $500 million annually in sales of goods and services. More workers at plants mean more people who need lunches and more people with money to spend.

Disadvantages of Nuclear Energy

Environmental Impact

One of the biggest issues is the environmental impact in relation to uranium. The process of mining and refining uranium hasn't been clean. Actually transporting nuclear fuel to and from plants involves a pollution hazard. Also, once the fuel is used, you can't simply take it to the landfill – it's radioactive and dangerous.

Radioactive Waste Disposal

As a rule, a nuclear power plant creates 20 metric tons of nuclear fuel per year, and with that comes a lot of nuclear waste. When you consider each nuclear plant on Earth, you will find that number jumps to approximately 2,000 metric tons a year.

The greater part of this waste transmits radiation and high temperature, implying that it will inevitably consume any compartment that holds it. It can also cause damage to living things in and around the plants.

Nuclear power plants create a lot of low-level radioactive waste as transmitted parts and supplies. Over time, used nuclear fuel decays to safe radioactive levels, however, this takes a countless number of years. Even low-level radioactive waste takes hundreds of years to achieve adequate levels of safety.

Nuclear Accidents

The accident in Three Mile Island in 1979, the Chernobyl accident that occurred on 26 April 1986 at the Chernobyl Nuclear Power Plant in Ukraine, was the worst nuclear accident in history. Then there was another accident that happened recently in Fukushima in Japan in 2011. Although the casualties were not that high, it caused serious environmental concerns. Its harmful effects on humans and ecology can still be seen today.

Despite all the safety measures in place in these nuclear plants, different factors caused them to go into meltdown causing devastating effects for the environment and for local inhabitants who had to leave the affected areas. The radioactive waste produced can pose serious health effects on the lives of people as well as the environment.

High Cost

The initial costs for building a nuclear power plant are steep. A recent virtual test reactor in the US estimate rose from $3.5bn to $6bn alongside huge extra costs to maintain the facility. South Africa scrapped plans to add 9.6GW of nuclear power to its energy mix due to the cost, which was estimated anywhere between $34-84bn.

At present, the nuclear business let waste cool for a considerable length of time before blending it with glass and putting it away in enormous cooled, solid structures. This waste must be kept up, observed and watched to keep the materials from falling into the wrong hands and causing problems.

These administrations and included materials cost cash and on top of the high expenses needed to put together a plant, which may make it less desirable to invest in. It requires permission from several international authorities, and it is normally opposed by the people who live in that region. The nuclear plants are cheap to run and produce inexpensive fuel, but the initial costs are huge.

Uranium is Finite

Typical renewable energy sources such as solar and wind are in infinite supply. Nuclear energy is not a renewable fuel source. Just like other sources of fuel, uranium is also finite and exists in a few of the countries. Uranium is in limited supply although currently abundant. There is still the risk of running out eventually.

Uranium has to be mined, synthesized, then activated to produce energy, and it's very expensive to go through this process. It produces a considerable amount of waste during all these activities and can result in environmental contamination and serious health effects, if not handled properly.

Hot Target for Militants

Nuclear energy has immense power. Today, nuclear energy is used to make weapons. If these weapons go into the wrong hands, that could be the end of this world. Nuclear power plants are a prime target for terrorism activities. Little lax in security can be brutal for humankind.

Fuel Availability

Unlike fossil fuels that are available to most of the countries, uranium is a very scarce resource and exists in only a few of the countries. Permissions of several international authorities are required before someone can even think of building a nuclear power plant.

DOE and its national labs are working with industry to develop new reactors and fuels that will increase the overall performance of these technologies and reduce the amount of nuclear waste that is produced. It also works to provide accurate, fact-based information about nuclear energy through its social media and STEM outreach efforts to educate the public on the benefits of nuclear energy.

Nuclear power has been used as a source of energy since the 1950s. Over the years, the source of energy has grown and there are about 440 nuclear power reactors in the world, accounting for about 10% of the planet's electricity. Nuclear energy is the energy contained within the core or nucleus of an atom. When these atoms split during nuclear fission, they release energy. The energy released can power homes, industries, schools, hospitals and much more.

Although the energy itself, produced by the nuclear power plants is renewable, the fuel used and required, is not. Uranium is the preferred fuel for nuclear fission in nuclear power plants. However, only one certain type of uranium (U-235) is suitable for the process. Uranium is a very common metal in the world, but the required isotope is comparatively rare to find, making it a non-renewable fuel for the renewable energy it produces.

The steam produced by the plant and used to power the turbines and the generators is recycled, making the source of energy all the more renewable. The steam is cooled down in a cooling tower and turned back into the water, which can be used again in the nuclear fission process.

Nuclear fission involves splitting the large atomic nucleus of the uranium 235 into smaller nuclei, producing energy. Also being investigated in length is nuclear fusion, which involves the combining of two or more atomic nuclei to form one or more different atomic nuclei and subatomic particles, like neutrons and protons. These processes produce renewable energy, although the materials are not easily available.

Is Nuclear Energy Sustainable?

Overwhelmingly, nuclear energy is sustainable. Through the production of nuclear energy, the process produces clean energy and does not result in environmental pollution or the release of greenhouse gases. This makes the energy more sustainable for the environment than fossil fuels.

The only by-product of the process is excess steam, which can simply be recycled into the atmosphere as clean water vapour. The biggest challenge which makes nuclear energy unsustainable at this particular minute is the fact that the uranium required is not readily available and even if it is, there is a human timescale on when it will be used up entirely.

If the source of the uranium is changed, or if the science allows for nuclear fusion, we will be able to have a sustainable source of the raw material, thereby making the source of energy more sustainable. Spring Power and Gas estimates that the remaining amounts of uranium in the world can serve us for the next 80 years, making nuclear energy sustainable and available to some generations to come. If we can turn nuclear fusion into a reality, the source of energy would be sustainable and able to sustain us for the foreseeable future. Is Nuclear Energy Good or Bad for the Environment? Nuclear energy is good for the environment, although there are some exceptions.

It Protects Air Quality

Nuclear energy results in zero emissions, making it a clean source of energy. The process sees uranium atoms split, producing energy, which then spins turbines generating electricity without the harmful by-products emitted by fossil fuels.

The US avoided more than 476 million metric tons of carbon dioxide emission in 2019, equivalent to removing 100 million cars from the roads. As such, nuclear energy helps in cleaning the air by removing tons of harmful air pollutants that lead to acid rain, smog, cardiovascular and lung diseases.

It Produces Minimal Waste

Nuclear energy is extremely dense at the amount 1 million times greater than that of traditional energy sources, making the amount of nuclear fuel used not as much as one would think. The waste from nuclear power plants is minimal, as much of it can be reprocessed and recycled, although the US does not do it. As such, producing nuclear fuel does little in damaging the environment.

Nuclear Energy has a Small Footprint on the Land

Nuclear energy produces more electricity on less land than any other clean air source. A 1000-MW nuclear facility needs just about one square mile to operate.

However, wind farms require 360 times more land area, while solar farms require 75 times more land area, to produce a similar amount of electricity. This means you need more than 3 million solar panels or 430 wind turbines (a capacity factor not included) to produce an equal amount of energy as a typical commercial reactor.

It has a Higher Energy Density

Nuclear fission, the process by which nuclear energy is attained, produces much greater amounts of energy than burning fossil fuels like gas, coal and oil. Nuclear fission is more efficient than producing energy through traditional fossil fuel means and as a result, it requires less fuel to power a nuclear energy power plant. This, therefore, creates less waste and has a lesser impact on the environment and economy. However, Nuclear Energy has some Challenges with Regards to the Environment.

It Produces Radioactive Waste

After producing the energy required, the remaining materials are very radioactive and need to be stored securely so that they do not pollute the environment. Radioactive waste from nuclear energy production is incredibly dangerous, to land, human health and other players in the environment. It is, therefore, the greatest challenge of these plants, to store such radioactive waste, as it cannot be destroyed. As such, they seal it securely in containers and keep it underground.

Dangerous Accidents

Sometimes, some accidents affect these power plants and the securely stored nuclear waste, leading to dangerous result and long-term effects of radiation. The Chernobyl incident in April 1986, is considered the worst nuclear disaster in history and has since killed thousands in indirect consequences such as thyroid cancers as well as other cancers. In March 2011, an earthquake hit Japan, destroying the Fukushima Daichi nuclear power plant and resulting in radiation leaks, deaths, and other serious effects.

Effects on the Environment

Nuclear power plants have other effects on the environment, apart from the waste they produce. Mining the uranium is not done in an environmentally-friendly process and the open pits left behind after mining it, are dangerous for everyone.

The process also results in erosions, pollutes nearby sources of water, affects plants and crops and severely affects miners and their general health due to the increased exposure to the radiation during extraction and processing.

The Waste is Forever

Although the waste might not be a lot, it is permanent, dangerous and forever. It cannot be buried in landfills like any other waste and therefore requires technologies that manage it well.

Is Nuclear Energy Safe?

Nuclear energy is safe for use. The same might not be said about the materials used or produced, but the energy, in itself, is safe.

Safe for the Environment

Nuclear power is safe for the environment. Nuclear energy results in zero emissions, making it a clean source of energy. Nuclear energy helps in cleaning the air by removing tons of harmful air pollutants that lead to acid rain, smog, cardiovascular and lung diseases. The process and resulting energy results in minimal waste and has a minimal footprint on the land, requiring a small portion of land as compared to wind and solar energy production plants.

Regulation

Nuclear power is one of the most highly regulated industries in the world. In the UK for instance, it is regulated by the Independent Office for Nuclear Regulation and the Environment Agency or the Scottish Environment Protection Agency (SEPA).

As such, all safety concerns have to be taken into consideration before a nuclear power plant begins or continues operations. This is to ensure that the process is successful and no harmful effects are placed on the environment, adjacent lands and surrounding communities, not forgetting the workers in these plants.

It is Reliable

Solar and wind power relies on the weather, underperforming if the weather conditions are unfavorable. Hydroelectric power also requires sufficient flow of water in dams and if prolonged droughts persist, the same is affected. Nuclear energy has no constraints and does not depend on the weather.

Nuclear power plants are unaffected by external climatic conditions and can continue to produce predictable and steady energy all day. A nuclear power plant can produce energy non-stop for an entire year, allowing for a good return on investment as there are no delays in energy production. Can Nuclear Energy Replace Fossil Fuels? The answer is a yes, which should actually be insisted upon.

Availability of Raw Materials

Materials used to arrive at nuclear energy are more available today than fossil fuels. If we continue consuming fossil fuels and keep increasing our consumption as the world population grows, Ecotricity estimates that we will run out of oil by 2052, gas in 2060, and coal by 2088. With regards to uranium, we have reserves that can last another 80 years.

Additionally, if scientists can turn nuclear fusion into a reality, we would theoretically never run out of electricity ever again. Some countries such as India, China, and Russia are also working towards using the greener and more abundant thorium, in place of uranium, to power nuclear reactors.

Nuclear Energy is Better than Fossil Fuel Energy

Nuclear energy is cleaner and cost-effective, holds more potential, is more abundant and is essential towards minimizing carbon dioxide emissions. It is also the better option for nations without a natural supply of fossil fuels and additional research into nuclear energy will result in the production of valuable technologies. The greatest advantage nuclear energy holds over fossil fuel is the fact that it results in zero emissions and is therefore safe for the environment.

It has a Higher Energy Density

Nuclear fission produces much greater amounts of energy than simply burning fossil fuels like gas, coal and oil. Nuclear fission is nearly 8000 times more efficient than producing energy through traditional fossil fuel means. As a result, it requires less fuel to power a nuclear energy power plant, therefore creating less waste and having a lesser impact on the environment and economy.

The Respective Wastes

Nuclear energy produces little nuclear waste. It is estimated that 60 years of waste from all US-based nuclear power plants, would fit inside a two-story building covering one city block. In contrast, waste from one coal plant operating for just one year can fill 11,000 freight cars, and most coal ash is slightly radioactive. Nuclear energy is, therefore, a lesser evil in comparison to fossil fuel.

Nuclear Fuel

Nuclear fuel is the fuel that is used in a nuclear reactor to sustain a nuclear chain reaction. These fuels are fissile, and the most common nuclear fuels are the radioactive metals uranium-235 and plutonium-239. All processes involved in obtaining, refining, and using this fuel make up a cycle known as the nuclear fuel cycle.

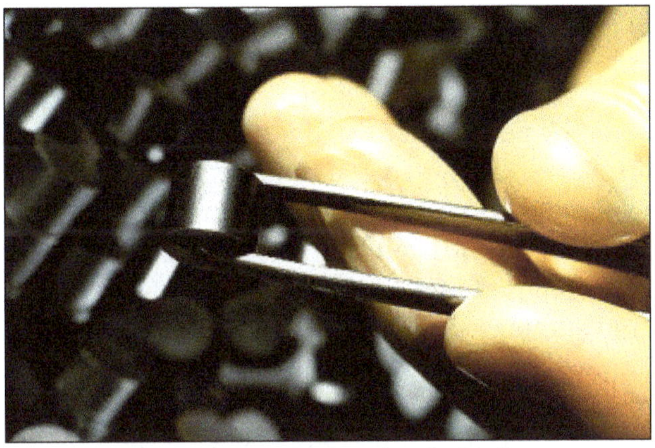

Figure: An enriched nuclear fuel pellet.

Uranium-235 is used as a fuel in different concentrations. Some reactors, such as the CANDU reactor, can use natural uranium with uranium-235 concentrations of only 0.7%, while other reactors require the uranium to be slightly enriched to levels of 3% to 5%. Plutonium-239 is produced and used in reactors (specifically fast breeder reactors) that contain significant amounts of uranium-238. It can also be recycled and used as a fuel in thermal reactors. Current research is being done to investigate how thorium-232 can be used as a fuel.

Production

Fuel fabrication plants are facilities that convert enriched uranium into fuel for nuclear reactors. For light water reactors, uranium is received from an enrichment plant in solid form. It is then converted into a gas and chemically converted into a uranium dioxide powder. This powder is then pressed into pellets and packed into fuel assemblies. A mixed oxide fuel can also be created when the uranium powder is packed along with plutonium oxide. The hazards present at fuel fabrication facilities—mainly chemical and radiological—are similar to the hazards at enrichment plants. These facilities generally pose a low risk to the public.

Use

When used in a reactor, the fuels used can have a variety of different forms a metal, an alloy, or some sort of oxide. Most nuclear reactors are fueled with a compound known as uranium dioxide.

This uranium dioxide is put together in a fuel assembly and inserted into the nuclear reactor—where it can stay for several months or up to a few years. While in the reactor the fuel undergoes nuclear fission and releases energy. This released energy is used to generate electricity. Neutrons released during the fission process allow for a fission chain reaction to occur, allowing energy to be generated continually. The fuel is removed from the reactor after large amounts of the fuel—whether it is uranium-235 or plutonium-239—have undergone fission. The "used" nuclear fuel is known as spent or irradiated fuel. After use, the fuel must be cooled for a few years as it is extremely hot.

The spent fuel is placed in large, deep pools of water that act as a coolant and a radiation shield. The coolant property allows the water to remove the decay heat and the shielding abilities protect workers from the radioactivity of the fuel. After cooling, the fuel can be re-purposed or sent to storage depending on regulations.

Fuel Assembly

Nuclear reactors are powered by powdered uranium dioxide that has been compressed into small pellets, shown in Figure. However, a power plant requires many of these pellets to run. Thus large numbers of these pellets are bundled into a fuel rod. A single uranium fuel pellet, only as big as a fingertip, contains as much energy as 481 cubic meters of natural gas, 807 kilograms of coal or 564 liters of oil. These rods are composed of numerous pellets of fissionable uranium fuel and can be several meters in length and about a centimeter in diameter. Then several of these rods, generally a dozen or more, are held together by strong metallic brackets in a fuel assembly. These rods are not bunched tightly together; rather there are several millimeters between each rod to allow coolant to flow between them. The tubes containing the pellets of uranium are generally composed of zirconium.

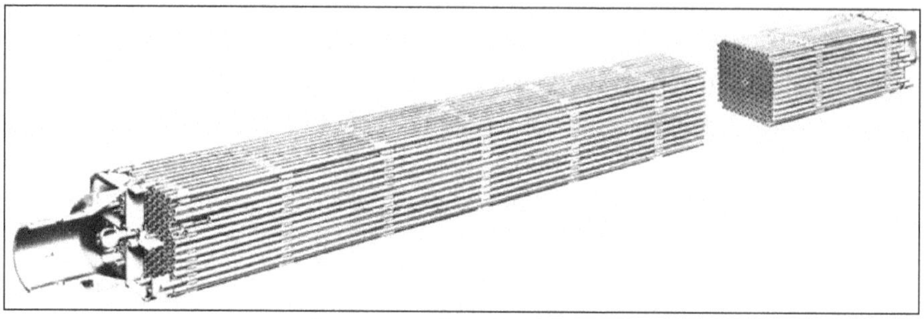

Figure: A nuclear fuel bundle.

Advantages and Disadvantages of Nuclear Fuels

Worldwide, there are extensive reserves of uranium left to mine. While nuclear fuel is not renewable, it is sustainable since there is so much of it. It will run out eventually, but not for centuries. Unlike fossil fuels, using nuclear fuels to produce energy does not directly produce carbon dioxide or sulfur dioxide. It should be mentioned that the processes of mining, transporting, and refining the fuel have carbon emissions associated with them, comparable to those of wind and solar power. Although the carbon footprint of using nuclear fuels is smaller, there are still disadvantages of using nuclear fuel. The waste, while a much lower volume must be handled very carefully because of its radioactivity. Nuclear fuels require far more complicated systems to extract their energy, which calls for greater regulation. These complex systems and regulation make for very long build times. In addition, public opinions on nuclear energy tend to be more negative than with other energy sources. The over-estimation of the dangers associated with releases of radioactive material is a significant issue, as large-scale nuclear incidents are rare.

Nuclear Fuel Cycle

The nuclear fuel cycle is the series of industrial processes which involve the production of electricity from uranium in nuclear power reactors. Uranium is a relatively common element that is found throughout the world. It is mined in a number of countries and must be processed before it can be used as fuel for a nuclear reactor. Fuel removed from a reactor, after it has reached the end of its useful life, can be reprocessed so that most is recycled for new fuel.

The various activities associated with the production of electricity from nuclear reactions are referred to collectively as the nuclear fuel cycle. The nuclear fuel cycle starts with the mining of uranium and ends with the disposal of nuclear waste. With the reprocessing of used fuel as an option for nuclear energy, the stages form a true cycle.

To prepare uranium for use in a nuclear reactor, it undergoes the steps of mining and milling, conversion, enrichment and fuel fabrication. These steps make up the 'front end' of the nuclear fuel cycle.

After uranium has spent about three years in a reactor to produce electricity, the used fuel may undergo a further series of steps including temporary storage, reprocessing, and recycling before the waste produced is disposed. Collectively these steps are known as the 'back end' of the fuel cycle.

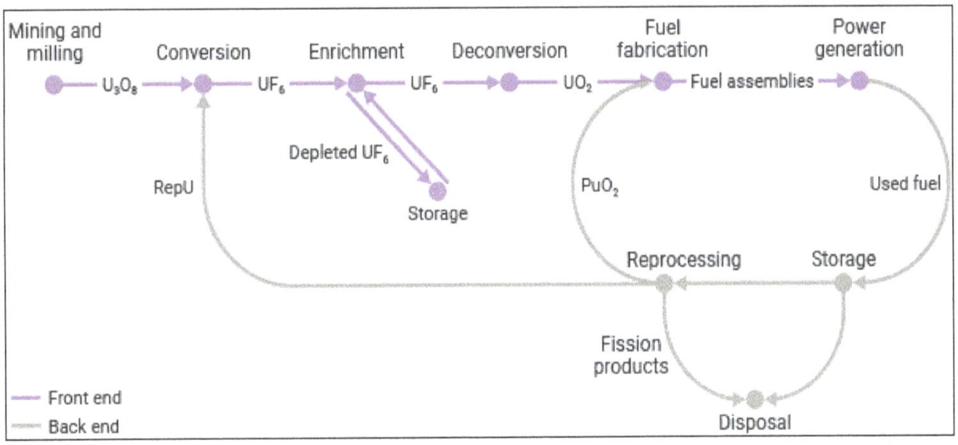

Figure: The nuclear fuel cycle.

Uranium

Uranium is a slightly radioactive metal that occurs throughout the Earth's crust. It is about 500 times more abundant than gold and about as common as tin. It is present in most rocks and soils as well as in many rivers and in sea water. It is, for example, found in concentrations of about four parts per million (ppm) in granite, which makes up 60% of the Earth's crust. In fertilisers, uranium concentration can be as high as 400 ppm (0.04%), and some coal deposits contain uranium at concentrations greater than 100 ppm (0.01%). Most of the radioactivity associated with uranium in nature is in fact due to other minerals derived from it by radioactive decay processes, and which are left behind in mining and milling.

There are a number of areas around the world where the concentration of uranium in the ground is sufficiently high that extraction of it for use as nuclear fuel is economically feasible. Such concentrations of minerals – including uranium – that can be extracted economically are referred to as ore.

Uranium Mining

The decision as to which mining method to use for a particular deposit is governed by the nature of the orebody, and safety and economic considerations. Both surface (generally open pit) and underground mining techniques are used to recover uranium ore. In general, open pit mining is used where deposits are close to the surface, and underground mining is typically used for deposits at depths greater than 120 m.

Since the walls of an open pit mine must be sloped to prevent collapse, the required holes are larger in size than the ore deposit it. As a result, the quantity of material that must be removed in order to access the ore may be very large. Underground mines have relatively small surface disturbance and the quantity of material that must be removed to access the ore is considerably less than in the case of an open pit mine. Special precautions, consisting primarily of increased ventilation, are required in underground mines to protect against airborne radiation exposure.

An increasing proportion of the world's uranium now comes from in situ leach (ISL) mining, where oxygenated groundwater is circulated through a very porous ore body to dissolve the uranium oxide and bring it to the surface. ISL may use either weak acid or alkaline solutions to keep the uranium in solution. The leached uranium oxide is then recovered from the solution as in a conventional mill.

Uranium Milling

Milling, which is generally carried out close to a uranium mine, extracts the uranium from the ore (or ISL leachate). Most mining facilities include a mill, although where mines are close together, one central mill may process the ore from several mines.

In a mill, the ore is crushed and ground to a fine slurry which is leached in sulfuric acid (or sometimes a strong alkaline solution) to allow the separation of uranium from the waste rock. It is then recovered from solution and precipitated as uranium oxide (U_3O_8) concentrate. After drying, and usually heating, it is packed in 200-litre drums as a concentrate, sometimes referred to as 'yellowcake' (though it is usually khaki in color). The U_3O_8 concentrate typically contains more than 80% uranium. The original ore, by comparison, may contain as little as 0.1% uranium.

U_3O_8 is the uranium product which is sold. About 200 tonnes is required to keep a large (1000 MWe) nuclear power reactor generating electricity for one year.

The remainder of the ore, containing most of the radioactivity and nearly all the rock material, becomes tailings, which are emplaced in engineered facilities near the mine (often in a mined out pit). Tailings need to be isolated from the environment because they contain long-lived radioactive materials in low concentrations and can contain toxic materials such as heavy metals. However, the total quantity of radioactive elements is less than in the original ore, and their collective radioactivity will be much shorter-lived.

Conversion and Enrichment

The uranium oxide product of a uranium mill is not directly usable as fuel for a nuclear reactor – additional processing is required. Only 0.7% of natural uranium is 'fissile', or capable of undergoing fission, the process by which energy is produced in a nuclear reactor. The form (or isotope) of uranium which is fissile is uranium-235 (U-235). The remainder is uranium-238 (U-238).

For most kinds of reactor, the concentration of the fissile U-235 isotope needs to be increased – typically to between 3.5% and 5% U-235. Isotope separation, the physical process to concentrate (or 'enrich') one isotope relative to others, requires the uranium to be in a gaseous form. The uranium oxide is first refined to uranium dioxide, which can be used as the fuel for those types of reactors that do not require enriched uranium. The balance – most of the uranium oxide – is then converted into uranium hexafluoride, which is a gas at relatively low temperatures. The uranium hexafluoride is then drained into 14-tonne cylinders where it solidifies. These strong metal containers are shipped to the enrichment plant.

The enrichment process separates gaseous uranium hexafluoride into two streams: one being enriched to the required level and known as low-enriched uranium; the other stream is progressively depleted in U-235 and is called 'tails', or simply depleted uranium.

The main enrichment process in commercial plants uses centrifuges, with thousands of rapidly-spinning vertical tubes. As they spin, the physical properties of molecules, specifically the 1% mass difference between the two uranium isotopes, cause them to separate. A laser enrichment process is in the final stage of development.

The product of this stage of the nuclear fuel cycle is enriched uranium hexafluoride, which is reconverted to produce enriched uranium oxide. Up to this point in the cycle the fuel material can be considered fungible (though enrichment levels vary), but fuel fabrication involves very specific design.

Fuel Fabrication

Reactor fuel is generally in the form of ceramic pellets. These are formed from pressed uranium oxide (UO2), which is sintered (baked) at a high temperature (over 1400°C). The pellets are then encased in metal tubes to form fuel rods, which are arranged into a fuel assembly ready for introduction into a reactor. The dimensions of the fuel pellets and other components of the fuel assembly are precisely controlled to ensure consistency in the characteristics and behaviour of the fuel.

In a fuel fabrication plant great care is taken with the size and shape of processing vessels to avoid criticality (a limited chain reaction releasing radiation). With low-enriched fuel criticality is very unlikely, but in plants handling special fuels for research reactors this is a vital safety consideration. Some 27 tonnes of fresh enriched fuel is required each year by a 1000 MWe reactor.

Power Generation and Burn-up

Several hundred fuel assemblies make up the core of a reactor. For a reactor with an output of

1000 MWe, the core would contain about 75 tonnes of low-enriched uranium. In the reactor core the U-235 isotope fissions or splits, producing a lot of heat in a continuous process called a chain reaction. The process depends on the presence of a moderator such as water or graphite, and is fully controlled.

Some of the U-238 in the reactor core is turned into plutonium and about half of this is also fissioned, providing about one-third of the reactor's energy output (or more than half in CANDU reactors).

As in fossil-fuel burning electricity generating plants, the heat is used to produce steam to drive a turbine and an electric generator. Through this process, a 1000 MWe unit provides over 8 billion kilowatt hours (8 TWh) of electricity in one year.

To maintain efficient reactor performance, about one-third of the spent fuel is removed every year or 18 months, to be replaced with fresh fuel. The length of fuel cycle is correlated with the use of burnable absorbers in the fuel, allowing higher burn-up.

Typically, some 44 million kilowatt-hours of electricity are produced from one tonne of natural uranium. The production of this amount of electrical power from fossil fuels would require the burning of over 20,000 tonnes of coal or 8.5 million cubic metres of gas.

An issue in operating reactors, and hence specifying the fuel for them, is fuel burn-up. Fuel burn-up is measured in gigawatt-days (thermal) per tonne and its potential is proportional to the level of enrichment. To date a limiting factor has been the physical robustness of fuel assemblies, and hence burn-up levels have been limited to about 40 GWd/t, requiring only around 4% enrichment. With the advancement of equipment and fuel assemblies, 55 GWd/t is now possible (with 5% enrichment), and 70 GWd/t is in sight (though this would require 6% enrichment). The benefit of increased burn-up is that operation cycles can be longer – around 24 months – and the number of fuel assemblies discharged as used fuel can be reduced by one third. Associated fuel cycle cost is expected to be reduced by about 20%. In CANDU reactors using natural uranium, burn-up is much less, about 7.5 GWd/t, but in terms of efficiency this is equivalent to almost 50 GWd/t for enriched fuel.

Burn-up in GWd/t is the conventional measure for oxide fuels, and 60 GWd/t U is equivalent to about 6.5 atomic percent burn-up (*i.e.* about 6.5% of the original uranium atoms are burned directly or indirectly via transformation to fissile plutonium). (With metal fuels, the atomic percent metric is used, and a new light water reactor metal fuel is targeting 21 atomic percent burns-up when it is deployed in 2020s).

As with coal-fired power stations, about two thirds of the heat produced is released, either to a large volume of water (from the sea or large river, heating it a few degrees) or to a relatively smaller volume of water in cooling towers, using evaporative cooling (latent heat of vapourisation).

Used Fuel

With time, the concentration of fission fragments and heavy elements in the fuel will increase to the point where it is no longer practical to continue using it. So after 18-36 months the used fuel is removed from the reactor. The amount of energy that is produced from a fuel assembly varies with the type of reactor and the policy of the reactor operator. Used fuel will typically have about 1.0% U-235 and 0.6% fissile plutonium (almost 1% Pu total), with around 95% U-238. The balance, about 3%, is fission products and minor actinides.

When removed from a reactor, the fuel will be emitting both radiation, principally from the fission fragments, and heat. It is unloaded into a storage pond immediately adjacent to the reactor to allow the radiation levels to decrease. In the ponds, the water shields the radiation and absorbs the heat, which is removed by circulating the water through external heat exchangers. Used fuel is held in such pools for several months and sometimes many years. It may then be transferred to naturally-ventilated dry storage, generally on site.

Depending on the policies of particular countries, some used fuel may be transferred to central storage facilities. Whilst there is a clear incentive for interim storage, used fuel must ultimately either be reprocessed in order to recycle most of it, or prepared for permanent disposal. The longer it is stored, the easier it is to handle, due to decay of radioactivity. There are two options for used fuel:

- Reprocessing to recover and recycle the usable portion of it.
- Long-term storage and final disposal without reprocessing.

Reprocessing

Used fuel still contains about 96% of its original uranium, of which the fissionable U-235 content has been reduced to less than 1%. About 3% of the used fuel comprises waste products and the remaining 1% is plutonium (Pu) produced while the fuel was in the reactor.

Reprocessing separates uranium and plutonium from waste products (and from the fuel assembly cladding) by cutting up the fuel rods and dissolving them in acid to separate the various materials. It enables recycling of the uranium and plutonium into fresh fuel, and produces a significantly reduced amount of waste (compared with treating all used fuel as waste). The remaining 3% of high-level radioactive waste (some 750 kg per year from a 1000 MWe reactor) can be stored in liquid form and subsequently solidified.

Uranium and Plutonium Recycling

The uranium recovered from reprocessing, which typically contains a slightly higher concentration of U-235 than occurs in nature, can be reused as fuel after conversion and enrichment. The plutonium can be directly made into mixed oxide (MOX) fuel, in which uranium and plutonium oxides are combined. In reactors that use MOX fuel, plutonium substitutes for the U-235 in normal uranium oxide fuel. According to Areva, about eight fuel assemblies reprocessed can yield one MOX fuel assembly, two-thirds of an enriched uranium fuel assembly, and about three tonnes of depleted uranium (enrichment tails) plus about 150 kg of waste. It avoids the need to purchase about 12 tonnes of natural uranium from a mine. Another way of recycling plutonium and uranium from reprocessing is Russia's REMIX, not yet commercialized. Here a non-separated mix of both has some low-enriched uranium (17% U-235) added, to produce fuel with about 1% Pu-239 and 4% U-235.Apart from the incidental transformation of U-238 into plutonium in a normal reactor, U-238 is not usable in today's fuel cycle. However, in a fast neutron reactor it is fissionable, as well as (more importantly) giving rise to plutonium, and is therefore potentially valuable. Increasingly, today's used fuel is being seen as a future resource rather than a waste.

Waste

Waste from the nuclear fuel cycle is categorised as high-, medium- or low-level based on the amount of radiation that it emits. This waste comes from a number of sources and includes:

- Low-level waste produced at all stages of the fuel cycle.
- Intermediate-level waste produced during reactor operation and by reprocessing.
- High-level waste, which is waste containing the highly-radioactive fission products separated in reprocessing, and in many countries, the used fuel itself. Separated high-level waste also contains long-lived transuranic elements.

After reprocessing, the liquid high-level waste can be calcined (heated strongly) to produce a dry powder, which is incorporated into borosilicate (Pyrex) glass to immobilise it. The glass is then poured into stainless steel canisters, each holding 400 kg of glass. A year's waste from a 1000 MWe reactor is contained in five tonnes of such glass, or about 12 canisters 1.3 metres high and 0.4 metres in diameter. These can readily be transported and stored; with appropriate shielding.

The uranium enrichment process leads to the production of much 'depleted' uranium, in which the concentration of U-235 is significantly less than the 0.7% found in nature. Small quantities of this material, which is primarily U-238, are used in applications where it's very high density characteristics are required, including radiation shielding the production of MOX fuel. While U-238 is not fissile it is a low specific activity radioactive material and some precautions must, therefore, be taken in its storage or disposal.

Used Fuel and Separated Waste Final Disposal

At the present time, there are no disposal facilities (as opposed to storage facilities) in operation in which used fuel not destined for reprocessing, and the waste from reprocessing, can be placed. In either case the material is stored in a solid, stable waste form.

There is currently no pressing logistical need to establish such facilities, as the total volume of such waste is relatively small. Further, the longer it is stored the easier it is to handle, due to the progressive decrease of radioactivity.

There is also a reluctance to dispose of used fuel because it represents a significant energy resource which could be reprocessed at a later date to allow recycling of the uranium and plutonium. A number of countries are carrying out studies to determine the optimum approach to the disposal of used fuel and waste from reprocessing. The general consensus favors its placement into deep geological repositories, about 500 meters down, initially recoverable before being permanently sealed.

Other Sources of Nuclear Fuel

In the 1990s uranium mines gained a competitor – in many ways very welcome – as military uranium came onto the civil market under a US-Russian agreement. In the period to 2013 half of the uranium used for electricity in the USA came from Russian military stockpiles, and worldwide about one-sixth of the market was supplied thus. In August 2013, the final shipment of low enriched uranium from Russian TVEL's JSC Electrochemical Plant marked the completion of Russia's commitments under the agreement commonly referred to as the 'Megatons to Megawatts' programme.

Weapons-grade uranium in stockpiles built up during 1950s and 1960s has been enriched to more than 90% U-235 and must be diluted about 1:25 or 1:30 with depleted uranium (about 0.3% U-235) in order to be used for civil power purposes. Weapons-grade plutonium may also be used to make mixed oxide (MOX) fuel for use in ordinary reactors or in special reactors designed to 'burn' it for electricity. Another US-Russian agreement covers disposition of military plutonium from both countries into MOX fuel.

Material Balance in the Nuclear Fuel Cycle

The following figures may be regarded as typical for the annual operation of a 1000 MWe nuclear power reactor typical of many operating today, using 4.5% enriched fuel and with 45 GWd/t burn-up:

Mining	Anything from 20,000 to 400,000 tonnes of uranium ore.
Milling	249 tonnes of uranium oxide concentrate (which contains 211 tonnes of uranium).
Conversion	312 tonnes of uranium hexafluoride, UF_6 (with 211 tU).
Enrichment	35.9 tonnes of enriched UF_6 (containing 24.3 t enriched U @ 4.5%) – balance is 'tails' @ 0.22%.
Fuel fabrication	27.6 tonnes UO_2 (with 24.3 t enriched U).
Reactor operation	8760 million kWh (8.76 TWh) of electricity at 100% output, hence 24 tonnes of natural U per TWh.
Used fuel	27.6 tonnes containing 280 kg transuranics (mainly plutonium), 26 t uranium oxide (<1.0% U-235), 1 tonne fission products.

The following figures assume the annual operation of 1000 MWe of nuclear power reactor capacity such as in the new AP1000 or EPR, with 5% enriched fuel and higher (65 GWd/t) burn-up:

Mining	Anything from 20,000 to 400,000 tonnes of uranium ore.
Milling	192 tonnes of uranium oxide concentrate (which contains 163 tonnes of uranium).
Conversion	241 tonnes of uranium hexafluoride, UF_6 (with 163 tU).
Enrichment	25 tonnes of enriched UF_6 (containing 16.85 t enriched U) – balance is 'tails' @ 0.22%.
Fuel fabrication	19.1 tonnes UO_2 (with 16.85 t enriched U).
Reactor operation	8760 million kWh (8.76 TWh) of electricity at 100% output, hence 18.6 tonnes of natural U per TWh.

Used fuel	19.1 tonnes containing 200 kg transuranics (mainly plutonium), 18.3 t uranium oxide (<1.0% U-235), <0.6 t fission products.

Efficiencies on power plant operation and lower enrichment tails assays meant that uranium demand per unit capacity was falling, and the report's generic reactor fuel consumption was reduced from 175 tU per GWe per year at 0.30% tails assay (2011 report) to 163 tU per GWe per year at 0.25% tails assay. The corresponding U_3O_8 figures are 206 tonnes and 192 tonnes per GWe per year.

Nuclear Fuel Resources

The need to meet ever-growing energy demands in an environmentally sustainable manner has turned attention to the potential for nuclear energy to play an expanded role in future energy supply mixes. One of the key aspects in defining the sustainability of any energy source is the availability of fuel resources.

Uranium Availability

Traditionally, these resources are categorised based on their economic attractiveness and on the confidence in their existence.

Conventional Resources

The most readily accessible resources, i.e. resources that are known to exist and are inexpensive to exploit using conventional mining techniques, are classed as "known conventional resources". These resources are categorised into two sub-groups: reasonably assured resources (RAR) and estimated additional resources category I (EAR-I). Known conventional resources are normally reported in terms of the amount of uranium recoverable, taking into account mining and milling process losses. They are reported in cost categories of resources recoverable at less than USD 40/kilogram of uranium (kgU), less than USD 80/kgU and less than USD 130/kgU. It should be noted that the current market price for uranium is about USD 20-30/kgU.

Resource type	Estimate (1000t)
Known conventional resources Reasonably assured resources (RAR) Estimated additional resources cat. I (EAR-I)	2850 1080
Undiscovered conventional resources Estimated additional resources cat. II (EAR-II) Speculative resources (SR)	2330 9940
Secondary sources Commercial inventories Surplus defence inventories Re-enrichment	220 250 440 Sub-total 17110

Unconventional resources In phosphates In seawater	22000 4000000
	Total 4039110

Resources believed to exist and to be exploitable using conventional mining techniques, but not yet physically confirmed, are classed as "undiscovered conventional resources". These resources include estimated additional resources category II (EAR II), uranium resources that are expected to be located in well-defined geological trends of known ore deposits, or mineralised areas with known deposits; and speculative resources (SR), uranium resources that are thought to exist in geologically favourable, yet still unexplored areas. Almost all EAR-II and SR are reported as in situ resources that are without providing for mining and milling losses.

Unconventional Resources

There are additional resources classified as unconventional, in which uranium exists at very low grades, or can be recovered as a minor byproduct. Unconventional resources include the uranium contained in phosphate deposits and in seawater. The technology to recover the uranium from phosphates is mature and has been utilised in the past; it is only the high costs of recovery that limits the desirability of recovering these resources. Research has hinted that it is possible to tap the vast resources of uranium contained in the world's oceans. At present, only laboratoryscale quantities have been extracted and as yet the cost to extract uranium from seawater is estimated to be very high, approximately five to ten times the cost of conventionally mined uranium. This technology would require additional time and investment to bring to deployment. Given the current low cost of uranium, with low-cost resources sufficient for several decades at current demand rates, it is doubtful that any significant funding will be made available in the foreseeable future.

Secondary Sources

Secondary sources of uranium, though small compared with the resources described above, play a significant role in supplying current nuclear fuel requirements and are expected to continue to do so through the near future. Important secondary sources of uranium are:

- Inventories of previously mined uranium held by both government and commercial organisations. These include strategic stocks, pipeline inventory and excess stocks available to the market.
- Large inventories of previously mined uranium derived from military applications in both the United States and the Russian Federation are becoming available for commercial applications. Highly enriched uranium (HEU) and natural uranium held in various forms by the military sector could total a few years' supply of natural uranium equivalent. In addition, surplus plutonium is available which, converted to mixedoxide fuel, displaces the need for fresh uranium.
- Large inventories of depleted uranium, the byproduct of the uranium enrichment process, represent a major reserve of uranium that could displace primary production of uranium.

As of 1999, the depleted uranium stockpile of 1.2 million tU could provide up to 452 000 tU of equivalent natural uranium.

Table provides estimates of the various resource categories discussed above, as of the end of 2000. The table highlights that over 75% of the conventional uranium resources are in the undiscovered conventional resources category. Thus, as known conventional resources become exhausted, future production will have to come from new projects. Known conventional resources could last about 75 years at the 2000 demand rate. While this is some time into the future, low cost resources will deplete sooner. Because new projects take a long time to reach production level, timely development decisions will be needed to ensure that resources become available when required. For example, the McArthur River deposit in Canada was discovered in 1988 and began production in 1999; eleven years were needed to bring the mine into production. Similarly, the Cigar Lake deposit, also in Canada, was discovered in the early-1980s but production is not expected to start before 2005. Numerous factors can influence the time necessary to develop a new mine project, including regulatory processes, legal challenges and economic conditions.

Geographical Distribution and Security of Supply

Another aspect of resource availability is the distribution of that resource around the world. Oil and natural gas have a fairly limited geographical availability, with the Middle East and the Russian Federation controlling some 70% of world crude oil and natural gas reserves.4 conversely, the OECD share of known uranium resources is roughly 40%, about the same as its share of coal reserves. This is much higher than its share of oil reserves (about 7%) and natural gas reserves (roughly 12%). Looking to the future, any country with access to the sea could ultimately have access to the vast uranium resources of the planet's oceans.

In addition, OECD countries are self-sufficient in the essential services that turn the uranium raw material into the finished nuclear fuel, that is, conversion, enrichment and fuel fabrication. Thus, nuclear power plants, once constructed, provide a largely if not entirely domestic source of electricity.

Effect of Advanced Reactor Technology and Fuel Cycles

Advanced technology is particularly important to any discussion on the availability of nuclear fuel resources because of its potential to radically extend the resource base and increase the efficiency of use. Light water reactors make up almost 80% of the reactors in operation worldwide. Most use a "once-through" fuel cycle in which the uranium placed in the reactor to produce energy is removed after a period of time and treated as waste for disposal. In this once-through fuel cycle about 99% of the potential energy content in the nuclear fuel remains unused. Other reactors use recycled fuel which improves the energy yield. Recycling spent nuclear fuel in current plants can save up to approximately 10 to 15% of the initially mined uranium through the use of the remaining uranium and the plutonium created during the fission process in the original uranium fuel. Plutonium is extracted from the spent fuel and recycled in mixed-oxide (MOX) fuel, but the number of recycles is currently limited because of the buildup of undesirable isotopes. After a few cycles the fuel would have to be managed as a waste similar to the once-through cycle. At present only a single recycle is used.

Yet, with already identified advances in technology these resources can be extended so that nuclear fuel resources become virtually unlimited. The introduction and use of fast reactors would provide significant benefits over current thermal reactor technology:

- They permit the more efficient use of fertile materials as nuclear fuel, such as uranium-238 and thorium, thereby expanding the available resource base.
- In a breeder configuration, fast reactors can produce more fuel than they consume, with this fuel being recovered and used to produce energy.

For example, fast breeder reactors could use the depleted uranium tails discussed above that have already been refined and are in storage. This byproduct of uranium enrichment would therefore become a fuel resource available for producing energy, augmenting natural uranium resources.

Thorium, an abundant and widely dispersed natural resource can also be used as a fuel resource. Existing estimates of thorium resources should be considered conservative for two reasons:

- Data from China, central and Eastern Europe, and the former Soviet Union have not yet been published.
- Historically weak market demand has limited the exploration for thorium.

It is possible; even likely, that much more thorium exists than has been documented and would be discovered if concerted exploration became warranted. Implementation of these technological innovations remains for the future and will require considerable research and development effort and investment. Although no commercial deployment of fast reactors is yet planned, they remain a promising opportunity in respect of resource utilisation.

Table: Nuclear fuel resources useable in fast reactors.

Resource category Estimate (1000t)
Thorium reserves 2160
Additional thorium resources 2350
Depleted uranium tails 1200 (758)[3]

- Data from China, central and Eastern Europe, and the former Soviet Union not available. World Energy Assessment, United Nations Development Program, New York, 2000; German Federal Institute for Geosciences and Natural Resources data bank.
- At the end of 1999. Management of Depleted Uranium, OECD, Paris, 2001.
- If the depleted uranium tails are re-enriched to the extent possible, about 758 000 tU will remain with 0.06% 235U assay.

Reactor Fuel cycle	Potential energy production- conventional resources only(TWH)	Potential energy production- total resources (TWH)	Years at 1999 world nuclear electricity generation (conventional)	Years at 1999 world nuclear electricity generation (total)
Current fuel cycle (LWR, once-through)	827 000	21 200 000	326	8 350

Recycling fuel cycle (Pu only, one recycle)	930 000	23 900 000	366	9 410
Light water and fast reactor mixed with recycling	1 240 000	31 800 000	488	12 500
Pure fast reactor fuel cycle with recycling	26 000 000	630 000 000	10 000	250 000
Advanced thorium/ uranium fuel cycle with recycling	43 200 000	90 200 000	17 000	35 500

- For a fuller description of the technologies listed and their resource requirements.
- Conventional resources were calculated using known conventional and undiscovered conventional resources plus secondary sources, from Table, for a total of 17 110 000 tonnes.
- Total resources assumes conventional resources plus the phosphate resources, but only 10% of seawater uranium is recovered (400 million t), added together for a total of 439 110 000 t.
- Key World Energy Statistics, IEA, Paris, 2001. Total 1999 (latest year data available) electricity generation by nuclear power was 2 538 TWh, rounded to 2 540 TWh.
- Assumes use of conventional uranium resources plus 758 000 t of depleted uranium tails remaining after re-enrichment. vi. Assumes use of only the thorium reserves base of 2 160 000 t with a matching amount of depleted uranium.
- Assumes thorium reserves base plus resources of 4 510 000 t with a matching amount of depleted uranium. This also assumes that no further discoveries of thorium are made.

Longevity

With these resource estimates and technology options, for how long will we be able to produce nuclear energy? The answer depends on many factors, including levels of electricity demand. Table demonstrates how long estimated resources would be able to produce electricity for various nuclear reactor and fuel cycle technology options. Illustrative assumptions were made as to the resources consumed by each technology as well as the levels of demand. Nevertheless, even at generation rates ten times greater than current levels, there are sufficient resources for many centuries of energy production, and technology developments would greatly extend these periods.

Nuclear Power Plant

Nuclear power plants are a type of power plant that uses the process of nuclear fission in order to generate electricity. They do this by using nuclear reactors in combination with the Rankine cycle, where the heat generated by the reactor converts water into steam, which spins a turbine and a generator. Nuclear power provides the world with around 11% of its total electricity, with the largest producers being the United States and France.

Figure: The Darlington nuclear power plant in Ontario produces power from four 878 MW CANDU reactors.

Aside from the source of heat, nuclear power plants are very similar to coal-fired power plants. However, they require different safety measures since the use of nuclear fuel has vastly different properties from coal or other fossil fuels. They get their thermal power from splitting the nuclei of atoms in their reactor core, with uranium being the dominant choice of fuel in the world today. Thorium also has potential use in nuclear power production; however it is not currently in use.

Components and Operation

Nuclear Reactor

The reactor is a key component of a power plant, as it contains the fuel and its nuclear chain reaction, along with all of the nuclear waste products. The reactor is the heat source for the power plant, just like the boiler is for a coal plant. Uranium is the dominant nuclear fuel used in nuclear reactors, and its fission reactions are what produce the heat within a reactor. This heat is then transferred to the reactor's coolant, which provides heat to other parts of the nuclear power plant.

Besides their use in power generation, there are other types of nuclear reactors that are used for plutonium manufacturing, the propulsion of ships, aircraft and satellites, along with research and medical purposes. The power plant encompasses not just the reactor, but also cooling towers, turbines, generators, and various safety systems. The reactor is what makes it differ from other external heat engines.

Steam Generation

The production of steam is common among all nuclear power plants, but the way this is done varies immensely.

The most common power plants in the world use pressurized water reactors, which use two loops of circling water to produce steam. The first loop carries extremely hot liquid water to a heat exchanger, where water at a lower pressure is circulated. It then heats up and boils to steam, and can then be sent to the turbine section. Boiling water reactors, the second most common reactor in power generation, heat the water in the core directly to steam.

Figure: Steam turbine in a power plant.

Turbine and Generator

Once steam has been produced, it travels at high pressures and speeds through one or more turbines. These get up to extremely high speeds, causing the steam to lose energy, therefore, condensing back to cooler liquid water. The rotation of the turbines is used to spin an electric generator, which produces electricity that is sent out the the electrical grid.

Figure: Two cooling towers of a nuclear power plant.

Cooling Towers

Perhaps the most iconic symbol of a nuclear power plant is the cooling towers, seen in figure. They work to reject waste heat to the atmosphere by the transfer of heat from hot water (from the turbine section) to the cooler outside air. Hot water cools in contact with the air and a small portion, around 2%, evaporates and raises up through the top. Moreover, these plants do not release any carbon dioxide—the primary greenhouse gas that contributes to climate change.

Many nuclear power plants simply put the waste heat into a river, lake or ocean instead of having cooling towers. Many other power plants like coal-fired power plants have cooling towers or these large bodies of water as well. This similarity exists because the process of turning heat into electricity is almost identical between nuclear power plants and coal-fired power plants.

Efficiency

The efficiency of a nuclear power plant is determined similarly to other heat engines—since technically the plant is a large heat engine. The amount of electric power produced for each unit of thermal power gives the plant its thermal efficiency, and due to the second law of thermodynamics there is an upper limit to how efficient these plants can be.

Typical nuclear power plants achieve efficiencies around 33-37%, comparable to fossil fueled power plants. Higher temperature and more modern designs like the Generation IV nuclear reactors could potentially reach above 45% efficiency.

Nuclear Reactor

A nuclear reactor is a system used to initiate and contain a nuclear chain reaction, and they have many useful applications. These nuclear reactions produce thermal energy through either nuclear fission (in practice) or nuclear fusion (in development). Nuclear reactors are primarily used for the generation of electricity, however they can be used for propulsion in vehicles such as submarines or naval vessels, for production of useful isotopes or neutrons, and for research and training.

Nuclear reactors can be found all over the world. Used in nuclear power plants, fission reactors contribute about 11% of the world's total electricity. Although there are many different designs for fission reactors, most consist of the same components for their operation. The difference between each type of fission reactor comes from the different approaches used to satisfy these requirements.

Figure: The core of a nuclear research reactor.

How do they Work?

The basic operating principles of a nuclear reactor for the production of power are as follows: Nuclear chain reactions within the reactor produce heat, which is transferred to a coolant (most commonly light water), the coolant either boils to steam directly or heats another loop of water into steam, it then passes through a turbine which spins a generator, and produces electricity. Although the basic principles appear simple, the process is fairly complex.

Fuel

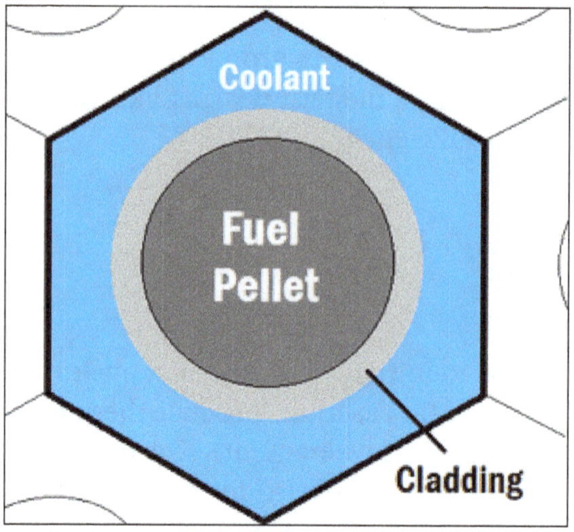

Figure: Fuel rods are surrounded by coolant to avoid overheating.

Nuclear reactors require the use of nuclear fuels, elements that can be readily altered and will release thermal energy. Uranium is the most common element used as a nuclear fuel, although thorium is also possible. The naturally occurring isotopes are found in countries such as Kazakhstan, Canada and Australia.

The uranium fuel is manufactured into small fuel pellets and are packed into fuel rods and surrounded by cladding to avoid leaking into the coolant. These fuel rods are assembled into a fuel bundle, as seen below. There can be hundreds of fuel bundles in a nuclear reactor, meaning there can be tens of thousands of fuel rods.

Nuclear Fuel

Fuel Enrichment

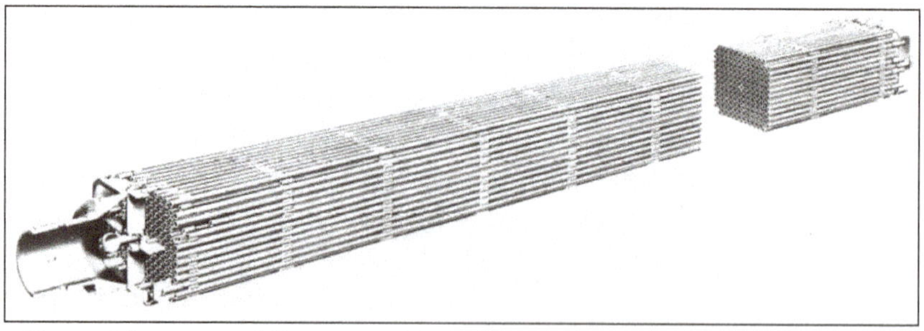

Figure: These assemblies are used in boiling and pressurized water reactors, and are packed into the core to produce heat in a nuclear reaction.

Not all nuclei of a given element are built exactly the same. An element is defined by the number of protons within the nucleus, and varying numbers of neutrons in the nucleus can cause it to behave differently. Natural uranium is mostly made up of uranium-238 (99.3%), with uranium-235 (0.7%) and very small amount of uranium-234 (0.0055%). Most reactors require a higher percentage of uranium-235 in order to sustain nuclear fission reactions, which can be done through uranium enrichment processes.

Moderator

Moderators are used to slow down the neutrons produced from fission. This is necessary because many nuclear fuels (uranium-235, for instance) require neutrons to be slow-moving in order to absorb them. Nuclei with low mass numbers are most effective at doing so, therefore materials like water or graphite are often used.

Most reactors use light water as a moderator, such as pressurized water reactors and boiling water reactors. Carbon works similarly and is used in reactors such as the RBMK. A third type of moderator used in CANDU reactors is heavy water, which is water composed of heavy hydrogen, called deuterium, rather than normal hydrogen.

The simulation below should help in visualizing how a moderator does its job: neutrons that are going too fast are absorbed by uranium-238 and do not yield fission (green) and the moderated neutrons are absorbed by uranium-235 which splits into smaller atoms and produces excess neutrons to continue the reaction (red).

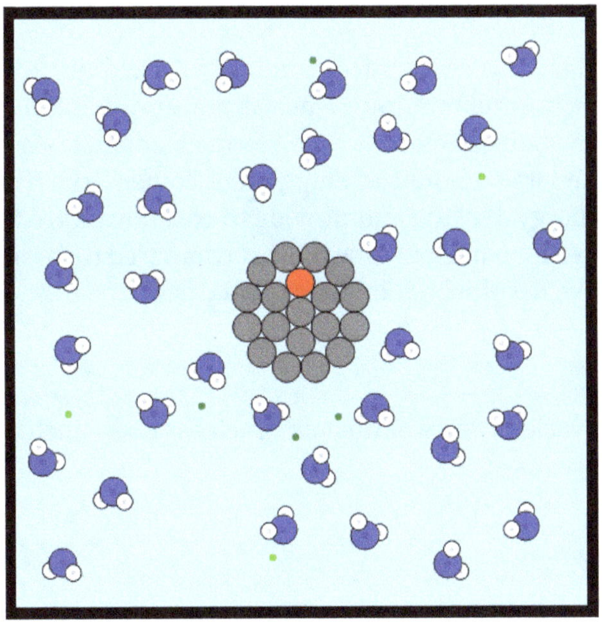

Coolant

The coolant, as its name implies, is used to remove heat from the core and move it to somewhere that it is useful. This keeps the fuel from overheating and melting down, at the same time as transferring the heat to water to make steam. Light water, heavy water, and various gases are the most common coolants for nuclear reactors. Coolants may also serve as the moderator, as is the case in many water-moderated reactors.

Control Rods

Control rods can be inserted into the reactor core to reduce the amount of fuel which undergoes fission reactions. The rods contain neutron absorbing atoms such as gadolinium or cadmium. By absorbing neutrons within the core, it prevents those neutrons from reacting with the fuel. Control rod movement can be used to adjust the number of reactions occurring at the core, or fully inserted to shut down the reactor completely.

Safety Systems

Safety systems are those in place to shut down the reactor and prevent radioactive material from being released. Some systems are passive, such as the dropping the control rods into the reactor core in CANDU reactors. The control rods are suspended above the core and held there by an electromagnet (a magnet which requires a constant supply of electricity to operate). In the event that a loss of power occurs, the control rods work to stop the reactions in the core. Strong containment buildings must also surround the reactor to prevent any radioactive leaks or external damage to the reactor.

Other safety systems require activation. An example of such a system is the release of large quantities of water to surround the reactor core. This provides cooling for the core to disperse the thermal energy and avoid a meltdown.

Economics

The building of nuclear reactors is economically intensive. The initial capital costs are high compared to fossil fuel plants with similar output. Nuclear power requires a high amount of additional safety, and is completely responsible for all possible nuclear waste. What makes nuclear power economically feasible is the large amount of energy that comes from a small volume of fuel. This relationship is known as energy density, and provides a cost advantage to using nuclear fuels. The cost of fuel is relatively lower for a nuclear power plant compared to fossil fuels. This is what makes nuclear reactors competitive despite high initial capital costs.

Reactor Types

Nuclear power consists of various types of nuclear reactors. These include:

- CANDU reactors,
- Fast breeder reactors,
- Thorium reactors,
- Boiling water reactors,
- Pressurized water reactors / Prismatic reactors,
- Molten salt reactors,
- Small modular reactors,
- Radioisotope thermal generators,

- Fusion reactors,
- RBMKs,
- Magnox reactors,
- Pebble bed reactors,
- Supercritical water cooled reactors,
- AES-2006/VVER-1000,
- VHTR,
- HTGR,
- Research reactors.

References

- How-nuclear-power-works, resources: ucsusa.org, Retrieved 13, April 2020
- Pros-and-cons-of-nuclear-energy: conserve-energy-future.com, Retrieved 10, March 2020
- Nuclear-Energy-Renewable: conserve-energy-future.com, Retrieved 27, July 2020
- Nuclear-fuel-resources-20-2: oecd-nea.org, Retrieved 12, May 2020
- Nuclear-fuel-cycle, information-library, nuclear-fuel-cycle-overview: world-nuclear.org, Retrieved 05, April 2020

Chapter 2

Water-cooled Reactors

Nuclear reactors that use water as a coolant and moderator are called water-cooled reactors. There are different types of water-cooled reactors. Boiling water reactor, pressurized water reactor, water-water energetic reactor, RBMK reactor, etc. are a few examples. This chapter deals with the various types of water-cooled reactors and examines them in a detailed manner for an extensive understanding of the subject.

Water cooled reactors have played a significant role in the commercial nuclear industry since its beginnings and currently account for more than 95 per cent of all operating civilian power reactors in the world. In addition, the majority of nuclear reactors under development and construction are water-cooled.

Water Cooled Reactors (WCRs) have been the cornerstone of the nuclear industry in the 20th century. Of the currently operating 442 reactors, 96 per cent are water-cooled. Many of these plants were originally licensed to operate for 40 years, but through advances in knowledge, the lives of these plants are now being extended to 60 years, with a potential to operate even longer. It can be expected that WCRs will continue to play an important role in the 21st century.

Light water reactors (LWRs) are the most common WCRs worldwide and are divided into two types: Pressurized Water Reactors (PWRs), which produce steam for the turbine in separate steam generators; and Boiling Water Reactors (BWRs), which use the steam produced inside the reactor core directly in the steam turbine. All LWRs require fuel that is enriched in the fissile isotope, U-235.

Heavy Water Reactors (HWRs) use "enriched" water, the molecules of which comprise hydrogen atoms that are made up to more than 99 per cent of deuterium, a heavier hydrogen isotope. This heavy water, used as a moderator, improves the overall neutron economy, allowing fuel to be used that does not require enrichment.

Many of the lessons learned from the past 50 years of WCR operation, including three major accidents, continue to be applied to the design and operation of existing and advanced WCRs. Recent advances in WCR technology include improvements in existing designs and the development of new designs, sharing the common goals of enhanced safety, a more efficient resource use and better economics. Another important aspect of WCR development is the design, testing and construction of small modular, factory-built integral pressurized water reactors.

While a number of advanced WCRs are currently under construction and available for deployment, innovations in WCR technology continue in the areas of safety, construction technology and economics. Safety systems designed into today's advanced WCRs have passive features that do not rely on electric power and include larger water inventories, allowing for coping times of days instead of hours in the event of unplanned conditions, such as an extended station blackout.

To improve plant thermal efficiencies and economics, research and development for Supercritical Water Reactors (SCWRs) is being pursued. Supercritical water exists at temperatures and pressures above its critical point, where the liquid and vapor states are indistinguishable. This water is commonly used in advanced coal, oil and gas-fired power plants. Plant efficiencies of SCWRs are expected to be around 1.3 times higher than conventional WCRs.

Supercritical Water Cooled Reactor

Supercritical water cooled reactor (SCWR) is a proposed Generation IV nuclear reactor which would function above the thermodynamic critical point of water. The critical point of water occurs when its temperature reaches 374 °C and its pressure is 22.1 MPa. At this point, water becomes neither a liquid nor a gas, but rather a distinct state with its own unique properties. The SCWR is one of three light water nuclear reactors, with the others being the pressurized water reactor and boiling water reactor. These are the most used reactors in the world for nuclear power generation.

The design of the SCWR follows a combination of the basic principles of boiling water reactors and supercritical coal plants, both of which have been used extensively and would help to reduce experimental hurdles experienced with new projects. The use of supercritical water would succeed in increasing the efficiency of the reactor to near 44%, compared to around 34% for current reactors—while potentially decreasing the capital cost by up to 30% due to smaller necessary containment and safety systems. This increase in efficiency would decrease the waste heat produced by the reactor by 35%, thereby making better use of the nuclear fuel in the reactor.

The main advantage of the SCWR would be improved economics because of the higher thermal efficiency and the potential for plant simplification. Research in the areas of safety, sustainability, and proliferation resistance and physical protection are being pursued by considering several design options using thermal and fast neutrons. The use of thorium for fuel also has potential in a SCWR below for a schematic of a SCWR.

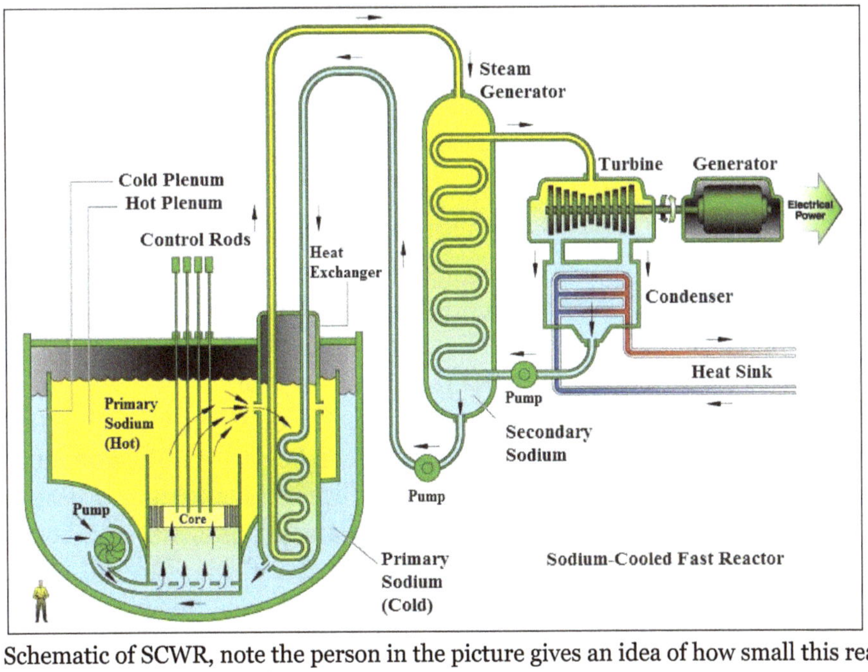

Figure: Schematic of SCWR, note the person in the picture gives an idea of how small this reactor is.

Supercritical Reactors Prototypes

Generally, the theoretical designs of SCWRs can be divided into two main categories: the pressure vessel concept proposed first by Japan (and more recently by The European Atomic Energy Community (Euratom) partnership), and pressure tube concept proposed by Canada (uncreatively called the Canadian-SCWR).

Japanese SCWR

The JSCWR uses neutrons that become slowed down, using light water as moderator and reactor coolant to become moderated neutrons. The JSCWR consists of a pressure-vessel type once-through reactor and a direct Rankine cycle system. The coolant goes into the reactor through inlet nozzles, heated up in the core and flows through outlet nozzles with no recirculation in the vessel. The core outlet coolant is directly delivered to the turbine system, and the feed water comes back directly from the feed water pumps. The electrical output is assumed to range from 600 MWe to 1700 MWe.

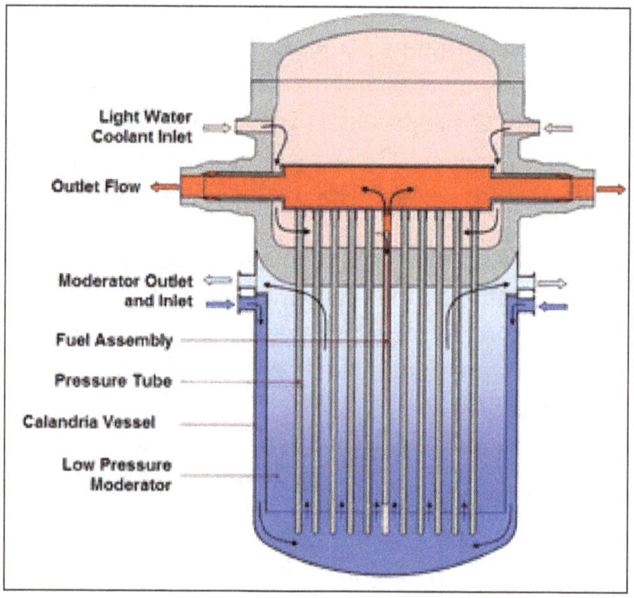

Figure: SCWR core design.

Canadian SCWR

The model of the Canadian-SCWR also slows down the neutrons (moderated neutrons) and maintains the CANDU modular design with fuel channels separating the light water coolant from the heavy water moderator. Operating conditions are set at a pressure of 25 MPa and the water comes out of the core 625 °C. This high outlet temperature would lead to an estimated thermal efficiency of 48%. The reheat option and co-generation capability can be achieved by advanced enhancement of the thermal efficiency. The supercritical steam is led directly to the high-pressure turbine, eliminating the need for steam generators which results in plant simplification and cost saving. Unlike the conventional CANDU design, Canadian SCWR has vertical pressure tubes to avoid thermal stratification. Unfortunately, this feature loses the ability to have on-line refueling. The electric power output is assumed to be about 1200 MWe.

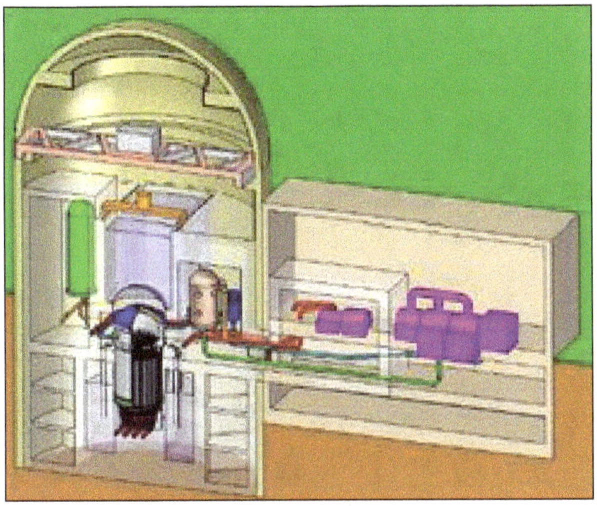

Figure: A layout of Canadian SCWR.

Euratom SCWR

The Euratom SCWR has a more complicated way of heating up the coolant by adding several more changes. This is to keep the hottest part of the cladding (covering for the nuclear fuel) below 650 °C. The once through steam cycle produces 1000 MWe, resulting in a net efficiency of 43.5%.

Difficulties

While many of the aspects of SCWR are attractive, further research needs to be done on the chemical and physical properties of materials. These materials need to be able to exist in the radiation field of a nuclear reactor at the same time as the somewhat extreme chemistry of supercritical water. Both radiation fields and supercritical water can lead to materials degrading over time. Further research is needed to make sure that these materials won't chemically corrode in this extreme environment. Research on the physical properties of these materials is needed to be certain that they won't crack or simply bend out of shape due to the high mechanical stress. Demonstration of passive safety systems must be completed and heat transfer models for the supercritical water also provide research hurdles which are necessary to surpass for future development of these systems.

Boiling Water Reactor

Boiling water reactors (BWR) are a type of nuclear reactor that use light water (ordinary water, as opposed to heavy water) as their coolant and neutron moderator. They are the second most used reactor for nuclear power generation in the world, next to the pressurized water reactor (PWR)—with 75 in operation as of 2018. Figure graphs the percentage of various types of reactors worldwide and the chart details the number of each.

With maximum operating temperatures of around 285 °C, the Carnot efficiency (maximum efficiency) of boiling water reactors in the production of power is 46%. However, the realistic efficiency of these reactors in power production is around 33-34%.

Figure: The Enrico Fermi Nuclear Generating Station in Newport, Michigan operates by use of a boiling water reactor.

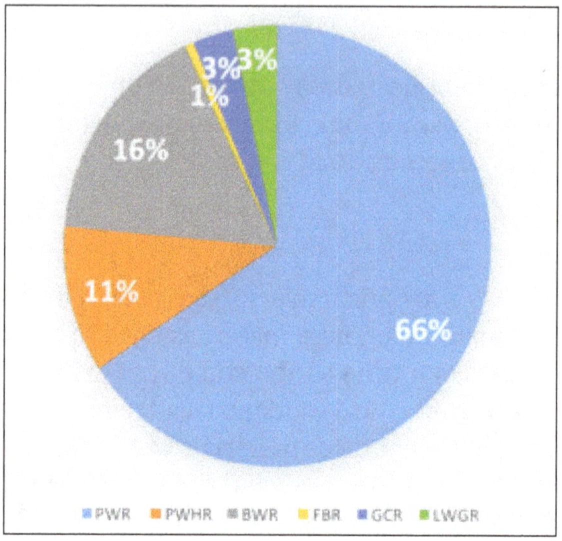

Figure: Percentage of Reactor types worldwide.

Type	Number of Reactors
PWR	297
BWR (Boiling Light-Water Cooled and Moderated Reactor)	75
PHWR (Pressurized Heavy-Water Moderated and Cooled Reactor)	49
LWGR (Light-Water Cooled, Graphite Moderated Reactor)	15
GCR (Gas Cooled, Graphite Moderated Reactor)	14
FBR (Fast Breeder Reactor)	3

Characteristics

Fuel

Boiling water reactors must use enriched uranium as their nuclear fuel, due to their use of light water. This is because light water absorbs too many neutrons to be used with natural uranium, so the fuel content of fissile Uranium-235 must be increased. This is done through uranium enrichment—which increases the concentration of Uranium-235 from 0.7% to around 4%.

The enriched uranium is packed into fuel rods, which are assembled into a fuel bundle, as seen in Figure. There are about 90-100 rods in each bundle, with up to 750 bundles in a reactor. This corresponds to nearly 140 tones' of uranium.

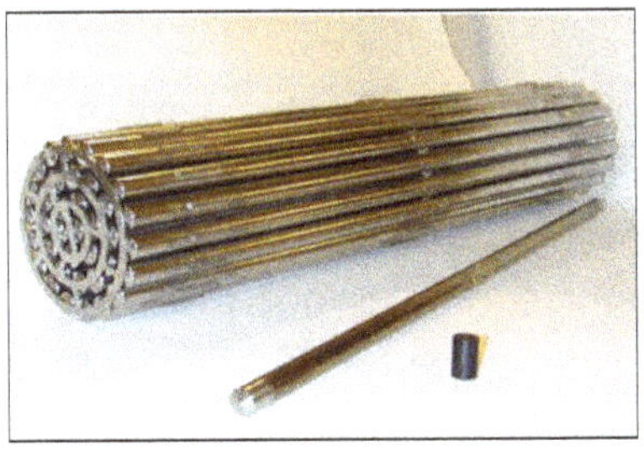

Figure: A nuclear fuel bundle.

Coolant and Moderator

Light water is used as the coolant and moderator for a boiling water reactor. Light water is much more abundant than heavy water, as it makes up 99.99% of natural water. The use of light water as a coolant is beneficial as it can cool down the fuel at the same time as it heats up to steam. The steam moves through the turbines which spin a generator to generate electricity. This simplifies the design of boiling water reactors, as there is only one primary loop needed which results in a lower cost of development.

Light water does not make as good of a moderator as heavy water or graphite due to its relatively high absorption of neutrons. However, its use as a moderator makes for an interesting and important safety feature; if there is a "loss of coolant accident: (LOCA), the moderator will also decline, causing the nuclear chain reaction to stop. In addition, if the moderating water overheats and turns to steam inside the bottom reactor core, there will be fewer moderators and therefore the chain reaction will stop.

Pressures

Boiling water reactors must operate at fairly high pressures in order for the water to perform adequately; at high pressures the water can remain in liquid form at temperatures much higher than its normal 100 °C boiling point. Therefore, boiling water reactors operate at around 7 MPa—around 70 times higher than atmospheric pressure.

Although very high, these pressures are still less than half of the pressures needed for pressurized water reactors. Since the pressures in boiling water reactors are less than those of pressurized ones, the specific power is less. This means that the boiling water reactor needs to be larger in order to provide the same power as a pressurized reactor, which results in higher costs. However, the walls do not need to be as thick as in PWRs, so the costs are essentially the same.

Water Flow

Since the BWR has only one primary loop, the very water that flows through the reactor core is the same one used to spin the turbines. In contrast, pressurized water reactors do not convert the hot water in the core to steam; therefore, two loops are necessary. This is the main difference between BWRs and PWRs.

Less water needs to flow through a BWR than through a PWR, however, there are complications because the water becomes radioactive in its circulation. This means that there must be extra shielding in the non-reactor parts of the power plant.

Control rods are placed in the bottom of the reactor, rather than the top as in PWRs. This is because in a boiling water reactor there is more steam at the top and placement of control rods in the steam portion would have a lesser effect than their placement in the water portion of the reactor. There is higher reactivity of the fuel in the water portion due to its moderation, so the control rods are needed to change the reactivity as deemed fit by the operators. Seen below is the flow of water through a boiling water reactor power plant, and its production of electricity.

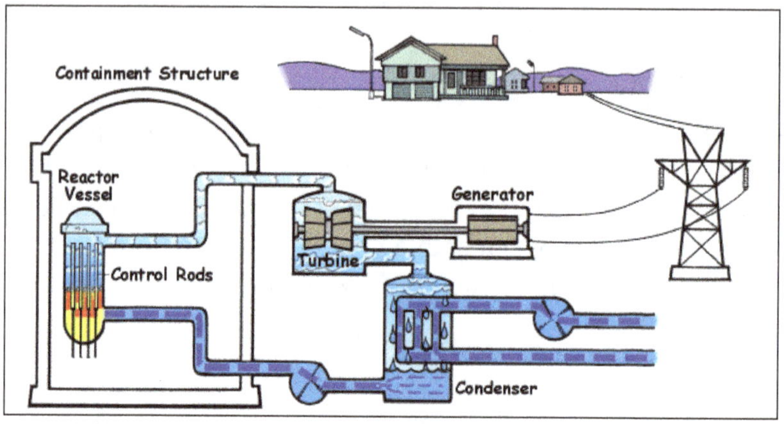

Figure: A boiling water nuclear reactor operates on a direct cycle, where the water from the core is also used as steam in the turbine section.

Pressurized Water Reactor

Pressurized Water Reactors were developed in the United States, as well as in Europe and the Soviet Union. Since most European development was as a result of licensing agreements with the United States, there is much commonality between the respective designs. Those developed in the Soviet Union were, however, different and with less rigorous safety provisions.

The principle of the pressurized water reactor is that the coolant circuit containing light water is pressurized to prevent boiling at the prevailing temperatures. The hot pressurized coolant is then used to generate steam in a secondary circuit at a lower pressure. The saturated steam so produced in the steam generators is fed directly to the steam turbines which are designed to operate with wet steam. Steam generators are of different designs with one type producing slightly superheated steam.

The fuel elements consist of many fuel rods extending the full height of the core and arranged in a square array. Guide tubes replace certain rods in alternate fuel elements. Neutron absorbing control rods are inserted into the guide tubes to regulate the heat production in the reactor. The coolant flows upward between the fuel rods removing the heat generated from the fission reaction. The coolant also serves as the moderator to reduce the energy of the neutrons so enhancing their capture in the fuel. Since light water does absorb some neutrons, slightly enriched fuel is required.

The reactor has to be shut down about once a year for refueling. However, only one third of the fuel is replaced each year and some moved within the reactor to ensure optimum fuel utilization. Generally, pressurized water reactors, with a relatively high power density in the reactor core, are compact and competitive in capital and operating costs. They are by far the most common type of reactors used for large scale commercial power generation.

The Pressurized Water Reactor, commonly known by its acronym PWR, is currently the most common type of reactor in commercial use for power production. These reactors were originally designed by Westinghouse Bettis Atomic Power Laboratory for military ship application and then by the Westinghouse Nuclear Power Division for commercial application. The first commercial PWR plant in the United States was Shippingport, which was located near Pittsburgh, Pennsylvania. Shippingport was a 60 MWe reactor that operated from 1957 to 1982 and later converted to the prototype LWBR. Following the 175 MWe Yankee-Rowe unit in 1961 and another in Italy, Westinghouse built reactors on a modular design (with from one to four steam generator loops) throughout the United States and under various licensing agreements with many countries including France, Germany, Japan, Sweden, and Spain. Moreover, Westinghouse was the vendor for the 1000 MWe Sizewall reactor, the first unit of a planned series of PWRs in the United Kingdom. Figure shows a simplified diagram of a typical PWR.

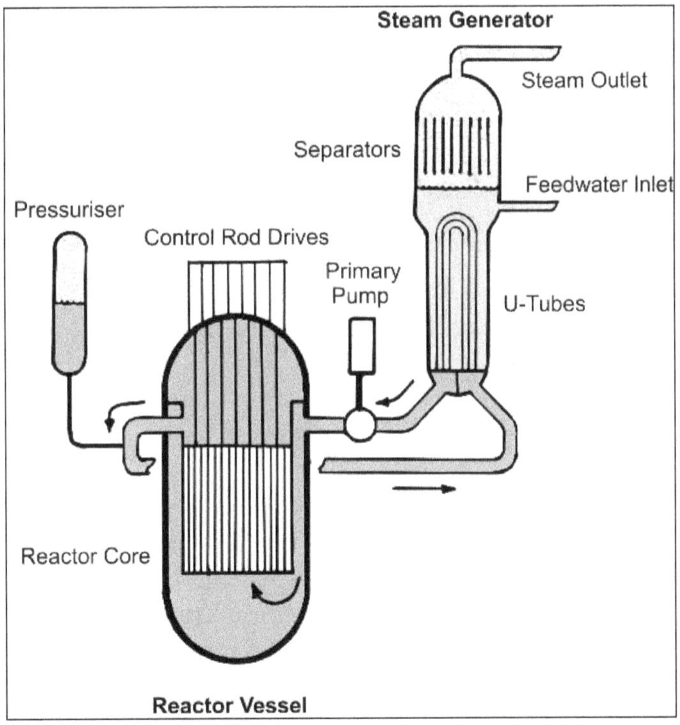

Figure: Diagrammatic cross-section of a typical PWR.

Other designs have been developed in the United States by Babcock & Wilcox (B&W) and Combustion Engineering (CE), in France by Framatome, in Germany by Siemens, Kraftwerk Union (KWU) and Brown Boveri (BBR), in Japan by Mitsubishi, and in the Soviet Union by Atommash in the late 1980s. Joint ventures have also been used with many other liaisons (especially for advanced "next generation" reactors). In addition to Westinghouse, Asea Brown Boveri-Combustion Engineering (ABB-CE), Framatome, Kraftwerk Union, Siemens, and Mitsubishi have typically built PWRs throughout the world. Babcock & Wilcox (B&W) built a PWR design power plant but used vertical once-through steam generators, rather than the U-tube design used by the rest of the suppliers. Framatome is now the world leader in PWR design and construction. It was a licensee to Westinghouse until 1984. Many renowned PWR designs have been built by Framotome. The 1450 MWe Chooz B1, an indigenous, state of the art PWR and the first unit in the N4-series, was scheduled for operation in 1992. Chooz B1 and Chooz B2 have been commercially operated since 1996. Framatome exports have gone to the Republic of South Africa, Belgium, South Korea, and the People's Republic of China.

The VVER is the Russian version of the commercial PWR developed from nuclear submarine and ice breaker experience. VVER is the abbreviation for Vodo-Vodyannoy Energeticheskiy Reactor which translates as "water-cooled water-moderated reactor". Soviet-designed nuclear power plants differ from Western nuclear power plants in many respects such as the VVERs use horizontal steam generators while western plants use two, three or four large vertical steam generators. Another main difference is that Soviet-designed reactors do not have a containment-structure (with exception of VVER1000 design). The initial 265 MWe unit at Novo Voronezh was followed with a series of 440 MWe and 1000 MWe units, with two of the former and one of the latter built at the same site. Although the VVER units were intended primarily for domestic and Eastern European use, two 445 MWe units have been in service at Loviisa in Finland since 1977 and 1981, respectively. In 1998, joint stock company Atomstroyexport (ASE) was established in order to promote the export of Russian-made products for nuclear power projects abroad.

Nowadays, the Russian nuclear plant designs have been developed based on the Russian norms and standards due to the international requirements. The latest export projects are two power units (2000 MW total) in China, a nuclear power plant of similar capacity in India and one power unit of 1000 MW in Iran.

Generally, in this chapter, PWR is used in reference to the designs originating in the United States and now widely utilized in Western countries and VVER to those originating in the Soviet Union and utilized within the Eastern Block.

Western Pressurized Water Reactor PWR

There are two major flow systems, the primary system and the secondary system, utilized to transport and convert the heat generated in the fuel into electrical power for industrial and residential use. The primary system transfers the heat from the fuel to the steam generator, where the secondary system begins. The steam formed in the steam generator is transferred by the secondary system to the main turbine generator, where its energy is converted into electricity. After passing through the turbine, the steam is routed to the main condenser. Cool water, passing through the tubes in the condenser from an external source, removes excess heat from the steam, allowing the steam to condense on the outside of the tubes. The condensate is then pumped back to the steam generator for reuse.

The primary system, also called the reactor coolant system, consists of 2, 3, or 4 cooling loops connected to the reactor, each containing a reactor coolant pump and steam generator. The systems built by the three main vendors consist of the same major components, but they are arranged in slightly different ways. For example, Westinghouse has built plants with 2, 3, or 4 loops, depending upon the power output of the plant. The Combustion Engineering plants and the Babcock & Wilcox plants only have two steam generators but they have four reactor coolant pumps. Figure shows the flow diagram for a three loop PWR with recirculating steam generators as in Figure. Boiling, other than minor bubbles arising from subcooled nucleate boiling, is suppressed by pressurization. Pressure is maintained by a pressurizer connected to the reactor coolant system. Pressure is maintained at approximately 15.5 MPa (2250 lbf in^{-2}) through a heater and spray system in the pressurizer. The primary coolant from the reactor is pumped to the steam generator and passes through its tubes. This water at a temperature of about 290 °C (554 °F) enters the reactor vessel through the inlet nozzles near the top. Water travels downward in the downcomer and then upward through the fuel assemblies. The water picks up heat and the temperature reaches about 310 °C to 332 °C (590 °F to 629 °F) as it exits the reactor.

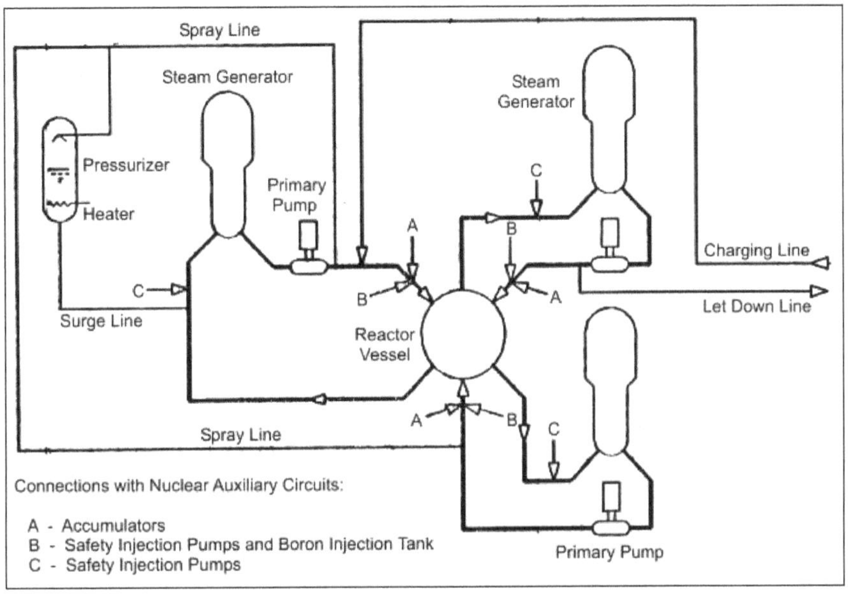

Figure: Primary circuit flow diagram.

In the secondary system, water is pumped into the steam generators and passes on the outside of the steam generator tubes, where the water is heated and converted to steam. The steam then passes through the main steam lines to the turbine, which is connected to and drives the electrical generator. The steam leaving the turbine condenses in a condenser. The condensate is then pumped by condensate pumps through low pressure feedwater heaters, then by the feedwater pumps through high pressure feedwater heaters and finally back to the steam generators. The condenser is maintained at a vacuum using either vacuum pumps or air ejectors. Cooling of the steam is provided by condenser cooling water pumped through the condenser by circulating water pumps which draw water from the ocean, lake, river, or cooling tower.

Soviet Pressurized Water Reactor VVER

There are three standard designs of VVER, namely, two different 6-loop 440 MW [440- 230 (older) and 440-213 (newer)] designs and a 4-loop 1000 MW design. As in the Western European and United States versions of the PWR, each reactor has primary coolant and secondary steam circuits. The primary coolant passes through the inside of the tubes in the steam generator. The reactor coolant pump circulates the water for cooling the reactor core. The system is pressurized to at least 15 MPa (2200 lbf in^{-2}) by a pressurizer which is connected to one of the reactor coolant loops. Spray valves and heaters in the pressurizer are used to control pressure in the coolant loop and reactor core. A major difference between Western designed PWRs and VVERs is that VVERs have horizontal steam generators. The older VVERs have isolation valves in the reactor coolant loops and accident localization compartments.

Secondary water, passing on the outside of the steam generator tubes, is heated and converted to steam. The steam passes to the turbine as in the PWR. The turbine drives the electrical generator. After being condensed in a condenser by cooling water, the condensate and feedwater are returned to the steam generator as in the PWR.

Major Differences between Soviet VVERs and Western PWRs

Soviet-designed nuclear power plants differ from Western-designed nuclear power plants in many respects, including plant instrumentation and controls, safety systems and fire protection systems. Although Soviet-designed plants, like Western-style plants, employ the design principle known in the West as "safety in depth", only the VVER-1000 design includes a containment structure as part of that principle.

In the unlikely event of a safety systems failure, plants designed on the "safety in depth" principle rely on a series of physical barriers to prevent the release of radioactive material to the environment. For typical Western-designed plants:

- The first barrier is the nuclear fuel itself, which is in the form of solid ceramic pellets. Most of the radioactive by-products of the fission process remain bound inside the fuel pellets.
- The fuel pellets are sealed in rods, made of special alloy and known as the cladding, to prevent the leakage of gaseous fission products.
- The fuel rods making up the reactor core are inside a large steel pressure vessel which, with the coolant loop pipe work and steam generators, forms a pressure boundary to contain the reactor coolant and any leaked radioactive material.
- At most plants, this vessel which serves to contain any fission products possibly released through the above barriers and associated coolant loop is enclosed in a large, leak-tight shell of steel plate and concrete known as the containment.

Most Soviet-designed reactors employ similar features as mentioned above, but only the VVER-1000 design has a containment structure like that of most nuclear power plants elsewhere in the world. Without this fourth level of protection, radioactive material could escape to the environment in the event of a serious accident involving the rupture of the reactor coolant circuit and failure of the fuel cladding.

At the end of 1994, about 90 commercial nuclear reactors of Soviet design were operating, under construction, and planned in Russia, Ukraine, Lithuania, Bulgaria, the Czech Republic, the Slovak Republic and Hungary. Of these commercial nuclear reactors, 63 were of the VVER type. A two-unit Soviet-designed nuclear plant in Finland was built using the VVER-440 Model 213 basic design, but was upgraded to include a Western instrumentation and control system and a containment structure. Recently, VVER-1000 designed nuclear power plants have been built in China, India and Iran.

Core Arrangement

Reactor Vessel

The major component of the reactor is the reactor vessel, in which is housed the core barrel, the reactor core, the upper internals package and all associated support and alignment devices.

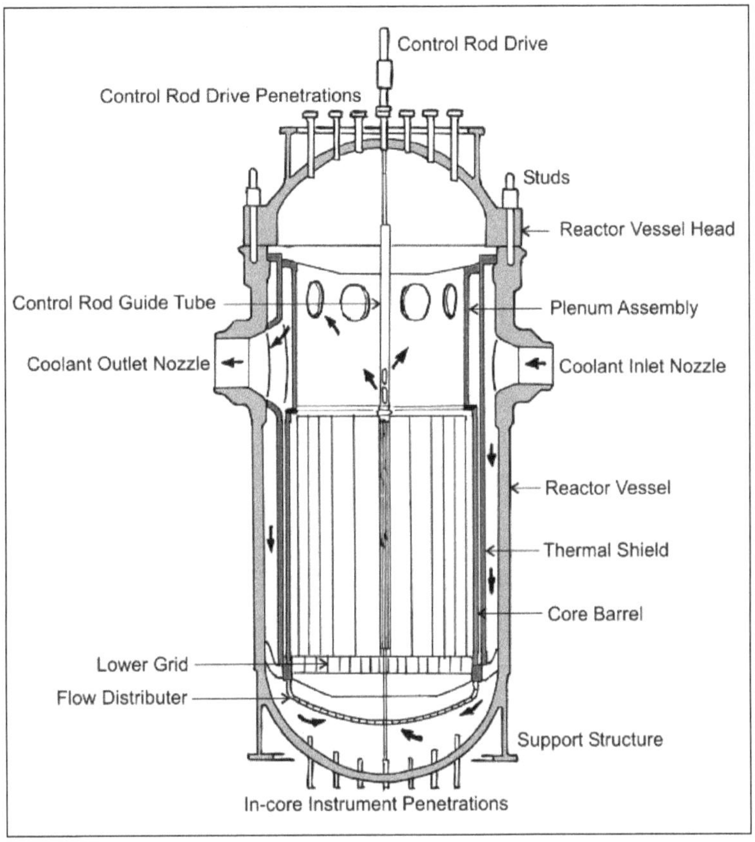

Figure: Cross-section of reactor vessel.

The reactor vessel, as shown in Figure, is a cylindrical vessel with a hemispherical bottom and a removable hemispherical head at the top. The head is removable to allow for the refueling of the reactor. There is one inlet (or cold leg) nozzle and one outlet (or hot leg) nozzle for each reactor coolant system loop. The reactor vessel is constructed of manganese molybdenum steel and all surfaces that come into contact with reactor coolant are clad with stainless steel to increase corrosion resistance.

The core barrel is fitted inside the reactor vessel and surrounds the fuel. It creates an annular channel for the incoming coolant and provides some radiation attenuation and hence protection for the reactor vessel. At the bottom of the core barrel, there is a lower core plate on which the fuel assemblies rest. The core barrel and all of the lower internals actually hang inside the reactor vessel from the internals support ledge. On the outside of the core barrel are irradiation specimen holders in which samples of the material used to manufacture the vessel are placed. At periodic intervals during the life of the reactor vessel, some of these samples are removed and tested to see how the radiation from the fuel has affected the strength of the material.

The upper internals package rests on top of the core. It contains guide columns to guide the control rods when they are pulled from or inserted into the core. The upper internals package also prevents the core from moving up during operation due to the force from the coolant flowing through the assemblies.

The coolant enters the vessel at the inlet nozzles. The core barrel forces the water to flow downward in the annulus between the reactor vessel and the core barrel. On reaching the bottom of the vessel, the flow is turned upward to flow through the fuel assemblies thus removing the heat produced by the fission process. The heated water enters the upper internals region, where it is routed out through the outlet nozzles. The flow is driven by the reactor coolant pumps in the cold leg of each coolant loop after passing through the steam generators. In the steam generators, the heat in the primary system is transferred to the secondary system and the reactor coolant returned to the reactor vessel.

Moderator and Coolant

All power reactor systems have a moderator which slows down neutrons via scattering interactions. An increase in temperature generally decreases both the density and the effectiveness for slowing down. The PWRs use light water as a moderator. It does, however, absorb neutrons to the extent that slightly enriched uranium fuel must be used to obtain a continuous fission chain reaction. The light water also serves as the coolant and circulates through the reactor. The moderator and coolant characteristics allow for a very compact reactor core. Since the fuel must be lumped and surrounded by both of moderator and coolant, it is formed into long thin vertical rods and arranged into a square array extending across the reactor core which thus takes the shape of a vertical cylinder. The spacing between the fuel rods is determined by the neutron slowing down distance in the light water. This heterogeneous arrangement also promotes effective heat transfer from the fuel to the coolant.

RBMK Reactor

The RBMK is an unusual reactor design, one of two to emerge in the Soviet Union in the 1970s. The design had several shortcomings, and was the design involved in the 1986 Chernobyl disaster. Major modifications have been made to RBMK reactors still operating.

The Soviet-designed RBMK (reaktor bolshoy moshchnosty kanalny, high-power channel reactor) is a water-cooled reactor with individual fuel channels and using graphite as its moderator. It is also known as the light water graphite reactor (LWGR). As with a boiling water reactor (BWR), water boils in the fuel channels (at about 6.9 MPa) and steam is separated above them in a single circuit. It was designed over 1964-66 and is very different from most other power reactors. Its precursors were an experimental 30 MWt (5 MWe) LWGR at Obninsk which started up in 1954, and two small prototype LWGR (AMB-100 & 200) units – which ran from 1964 and 1968 respectively.

(The ADE reactors at used for plutonium production are similar to the RBMK but with much lower power density and smaller fuel elements).

The combination of graphite moderator and water coolant is found in no other power reactors in the world. As the Chernobyl accident showed, several of the RBMK's design characteristics – in particular, the control rod design and a positive void coefficient – were unsafe. A number of significant design changes were made after the Chernobyl accident to address these problems.

A Light Water Graphite-moderated Reactor (LWGR/RBMK)

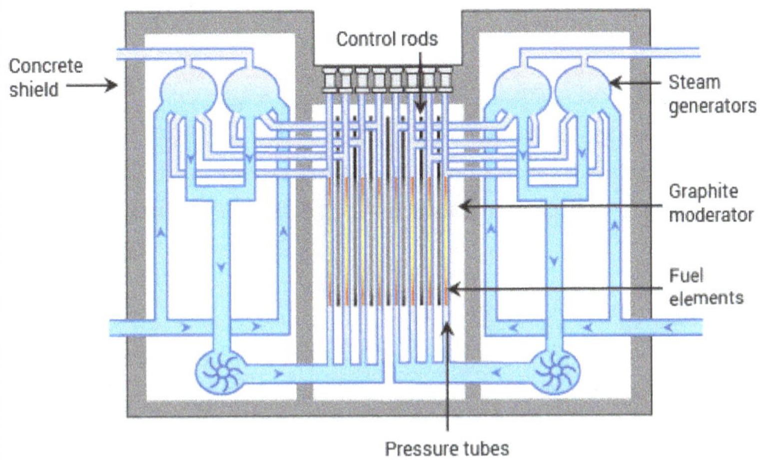

Features of the RBMK

Fuel

Pellets of slightly-enriched uranium oxide are enclosed in a zircaloy tube 3.65m long, forming a fuel rod. A set of 18 fuel rods is arranged cylindrically in a carriage to form a fuel assembly. Two of these end on end occupy each pressure tube.

Pressure Tubes

Within the reactor each fuel assembly is positioned in its own vertical pressure tube or channel about 7 m long. Each channel is individually cooled by pressurised water which is allowed to boil in the tube and emerges at about 290 °C.

Refuelling

When fuel channels are isolated, the fuel assemblies can be lifted into and out of the reactor, allowing fuel replenishment while the reactor is in operation.

Graphite Moderator

A series of graphite blocks surround, and hence separate, the pressure tubes. They act as a moderator to slow down the neutrons released during fission so that a continuous fission chain reaction can be maintained. Heat conduction between the blocks is enhanced by a mixture of helium and nitrogen gas.

Control Rods

Boron carbide control rods absorb neutrons to control the rate of fission. A few short rods, inserted upwards from the bottom of the core, even the distribution of power across the reactor. The main control rods are inserted from the top down and provide automatic, manual, or emergency control. The automatic rods are regulated by feedback from in-core detectors. If there is a deviation from normal operating parameters (*e.g.* increased reactor power level), the rods can be dropped into the core to reduce or stop reactor activity. A number of rods remain in the core during operation.

Coolant

Two separate water coolant loops each with four pumps circulate water through the pressure tubes to remove most of the heat from fission. There is also an emergency core cooling system which is designed to come into operation if either coolant circuit is interrupted.

Steam Separator

Each of the two loops has two steam drums, or separators, where steam from the heated coolant is fed to the turbine to produce electricity in the generator (each loop has a turbogenerator associated with it). The steam is then condensed and fed back into the circulating coolant.

Containment

There is no secure containment in the sense accepted in the West. The reactor core is located in a reinforced concrete lined cavity that acts as a radiation shield. The core sits on a heavy steel plate, with a 1000 tonne steel cover plate on the top. The extensions of the fuel channels penetrate the lower plate and the cover plate and are welded to each. The steam separators of the coolant systems are housed in their own concrete shields.

Positive Void Coefficient

The term 'positive void coefficient' is often associated with RBMK reactors. Reactors cooled by boiling water will contain a certain amount of steam in the core. Because water is both a more efficient coolant and a more effective neutron absorber than steam, a change in the proportion of steam bubbles, or 'voids', in the coolant will result in a change in core reactivity. The ratio of these changes is termed the void coefficient of reactivity. When the void coefficient is negative, an increase in steam will lead to a decrease in reactivity.

In those reactors where the same water circuit acts as both moderator and coolant, excess steam generation reduces the slowing of neutrons necessary to sustain the nuclear chain reaction. This leads to a reduction in power, and is a basic safety feature of most Western reactors.

In reactor designs where the moderator and coolant are of different materials, excess steam reduces the cooling of the reactor, but as the moderator remains intact the nuclear chain reaction continues. In some of these reactors, most notably the RBMK, the neutron absorbing properties of the cooling water are a significant factor in the operating characteristics. In such cases, the reduction in neutron absorption as a result of steam production, and the consequent presence of extra free neutrons, enhances the chain reaction. This leads to an increase in the reactivity of the system.

The void coefficient is only one contributor to the overall power coefficient of reactivity, but in RBMK reactors it is the dominant component, reflecting a high degree of dependence of reactivity on the steam content of the core. At the time of the accident at Chernobyl, the void coefficient of reactivity was so positive that it overwhelmed the other components of the power coefficient, and the power coefficient itself became positive. When the power began to increase, more steam was produced, which in turn led to an increase in power. The additional heat resulting from the increase in power raised the temperature in the cooling circuit and more steam was produced. More steam means less cooling and less neutron absorption, resulting in a rapid increase in power to around 100 times the reactor's rated capacity.

The value of the void coefficient is largely determined by the configuration of the reactor core. In RBMK reactors, an important factor affecting this is the operating reactivity margin.

Operating Reactivity Margin

Although the definition is not precise, the operating reactivity margin (ORM) is essentially the number of 'equivalent' control rods of nominal worth remaining in the reactor core. The operators at Chernobyl seemed to believe that safety criteria would be met so long as the lower limit for the ORM of 15 equivalent rods was adhered to, regardless of the actual configuration of the core. The operators were not aware of the 'positive scram' effect where, following a scram signal, the initial entry of the control rods actually added reactivity to the lower region of the core.

The ORM could have an extreme effect on the void coefficient of reactivity, as was the case for the core configuration of Chernobyl 4 in the run-up to the accident. Unacceptably large void coefficients were prevented for initial cores by increasing fuel enrichment levels, with the excess reactivity balanced by fixed absorbers. However, with increasing fuel burn-up, these absorbers could be removed to maintain the fuel irradiation levels - shifting the void coefficient in the positive direction and increasing the sensitivity of the coefficient to the extent of insertion of the control and protection rods.

Post-accident Changes to the RBMK

After the accident at Chernobyl, several measures were taken to improve the safety of RBMK plants. All operating RBMK reactors in the former Soviet Union had the following changes implemented to improve operating safety:

- Reduction of the void coefficient of reactivity.
- Improvement of the response efficiency of the emergency protection system.
- Introduction of calculation programs to provide an indication of the value of the operating reactivity margin (ORM, *i.e.* the effective number of control rods remaining in the core) in the control room.
- Prevention of the emergency safety systems from being bypassed while the reactor is operating.

- In order to ensure adequate subcooling at the core inlet, the avoidance of modes of operation that cause a reduction in the departure from nucleate boiling (DNB) ratio of the coolant at the reactor inlet.

Measures to reduce the void coefficient of reactivity were carried out by:

- The installation of 80-90 additional fixed absorbers in the core to inhibit operation at low power.
- Increasing the ORM from 26-30 rods (in steady state operational mode) to 43-48.
- An increase in fuel enrichment from 2% to 2.4%.

The increase in the number of fixed absorbers and the ORM reduced the value of the void coefficient of reactivity to +ß (where ß is the effective delayed neutron fraction). The additional absorbers require the use of higher fuel enrichment to compensate for the increased neutron absorption. The efficiency and speed of the emergency protection system was improved by implementing three independent retrofitting operations:

- Retrofitting of control rods with a design that does not give rise to water columns at the bottom of the channels.
- Scram (shut down) rod insertion time cut from 18 to 12 seconds.
- The installation of a fast-acting emergency protection (FAEP) system.

One of the most important post-accident changes to the RBMK was the retrofitting of the control rods. A graphite 'displacer' is attached to each end of the length of absorber of each rod (except for 12 rods used in automatic control). The lower displacer prevents coolant water from entering the space vacated as the rod is withdrawn, thus augmenting the reactivity worth of the rod. However, the dimensions of the rod and displacers were such that, with the rod fully withdrawn, the 4.5 m displacer sat centrally within the fuelled region of the core with 1.25 m of water at either end. On a scram signal, as the rod falls, the water at the lower part of the channel is replaced by the bottom of the graphite displacer, thus initially adding reactivity to the bottom part of the core. Following the Chernobyl accident, this 'positive scram' effect was mitigated by retrofitting the control rods so that, with the rods fully retracted, there would not be region containing water at the bottom of the core.

The FAEP system was designed so that 24 emergency protection control rods would insert negative reactivity of at least 2ß in less than 2.5 seconds. Tests in 1987-'88 at the Ignalina and Leningrad plants (the first RBMKs to be fitted with the new FAEP system) confirmed these characteristics. In addition to the above changes, several further modifications have been implemented at RBMK plants. These measures consist of:

- Replacement of the fuel channels at all units (except Smolensk 3).
- Replacement of the group distribution headers and addition of check valves.
- Improvements to the emergency core cooling systems.
- Improvements of the reactor cavity over-pressure protection systems.
- Replacement of the SKALA process computer.

Operating RBMK Plants

There are currently nine operating RBMKs, all of which are in Russia. One more was under construction in Russia (Kursk 5), but it is now cancelled. All operating RBMKs began operation between 1976 (Kursk 1) and 1990 (Smolensk 3). There are three distinct generations of RBMK reactors having significant differences with respect to their safety design features:

- The four first-generation units are Leningrad 1&2 (both now shutdown), and Kursk 1&2. They were designed and brought on line in the early-to-mid-1970s, before new standards on the design and construction of nuclear power plants, the OPB-82 General Safety Provisions, was introduced in the Soviet Union in 1982.
- Second-generation RBMKs, brought online since the late 1970s and early 1980s include Leningrad 3&4; Kursk 3&4; Ignalina 1 (now shutdown); and Smolensk 1&2. Ignalina 2 (now shutdown) had safety features beyond those of other second generation units. These units conform to the OPB-82 standards.
- After the Chernobyl accident, Soviet safety standards were revised again (OPB-88). One RBMK (Smolensk 3) has been built to these third-generation standards. Additional design changes were being incorporated in the construction of Kursk 5 before it was cancelled.

In 2006, Rosatom said it was considering operating lifetime extensions and uprating of its operating RBMK reactors. Following significant design modifications made after the Chernobyl accident, as well as extensive refurbishment including replacement of fuel channels, a 45-year operating lifetime is seen as realistic for the 1000 MWe-class units. In 2018, they provided 31% of Russia's nuclear-generated electricity.

In the Table below, the 'operating until' dates are the scheduled shutdown for these plants, with 15-year lifetime extensions in some cases. Lithuania, on the other hand, closed Ignalina 1&2 early as a condition for entry into the European Union.

Russia's long-term plans had earlier included the possibility of replacing the Leningrad units, at the end of their extended service life, by new MKER-1000 units. These are a modification of the RBMK design. The main differences are in the spacing of the graphite lattice in the core and the incorporation of passive safety systems.

RBMK Reactors

Operational dates are subject to licence extension.

Location	Unit	First power	Unit net capacity (MWe)	Status
Lithuania				
Ignalina	1	1983	1185 (originally 1300)	Closed 12/2004
	2	1987	1185 (originally 1300)	Closed 12/2009
Russia				
Kursk	1	1976	925	Operating until 2022
	2	1979	925	Operating until 2024
	3	1984	925	Operating until 2029
	4	1986	925	Operating until 2031

Location	Unit	First power	Unit net capacity (MWe)	Status
	5	-		Construction cancelled
Leningrad	1	1973	925	Closed 12/2018
	2	1975	925	Closed 11/2020
	3	1979	925	Operating until 2025
	4	1981	925	Operating until 2026
Smolensk	1	1983	925	Operating until 2028
	2	1985	925	Operating until 2030
	3	1990	925	Operating until 2050
Ukraine				
Chernobyl	1	1977	925	Closed 1996
	2	1978	925	Closed 1991
	3	1981	925	Closed 2000
	4	1983	925	Reactor destroyed April 1986
	5	-	925	Construction cancelled
	6	-	925	Construction cancelled

References

- Water-cooled-reactors: iaea.org, Retrieved 02, April 2020
- Supercritical-water-cooled-reactor: energyeducation.ca, Retrieved 09, July 2020
- Boiling-water-reactor: energyeducation.ca, Retrieved 13, May 2020
- Nuclear-power-reactors, information-library, nuclear-fuel-cycle: world-nuclear.org, Retrieved 30, June 2020

Chapter 3

Gas-cooled Reactors

Gas-cooled reactors are those nuclear reactors which use graphite as a neutron moderator and carbon dioxide as a coolant. They offer potential operational and safety benefits over water-cooled reactors. Magnox reactor, UNGG reactor, advanced gas-cooled reactor, high temperature cooled reactor and gas cooled fast reactor are a few gas-cooled reactors that are discussed thoroughly in this chapter.

GCRs offer potential operational and safety benefits over water-cooled reactors. A major operational motivation for considering this technology is improved energy conversion efficiency afforded by a higher reactor operating temperature. For instance, water-cooled reactors have a practical maximum temperature limit of about 350 °C, which yields a heat-to-electricity conversion efficiency of roughly 32–34%. By comparison, a GCR can operate at temperatures up to 800–850 °C and yield an energy conversion efficiency exceeding 40% when using conventional steam turbine equipment or as high as 50% when using more advanced gas turbine equipment.

From a safety perspective, GCRs typically use a lower core power density and a high heat capacity core, which helps limit fuel temperatures following a loss-of-coolant accident. Because they use a different fuel form and cladding, they avoid steam/zirconium cladding chemical reactions that can release explosive hydrogen gas under accident conditions in LWRs. Unlike traditional PWRs and BWRs, some GCR designs have the capability of being refueled during full-power operation, which provides some operational convenience and higher plant availability.

Several GCRs have been built and operated using either CO_2 or helium to cool the reactor core. Commercial GCRs use a graphite neutron moderator, which absorbs fewer neutrons than water-moderated reactors. The first-generation GCRs were built in the United Kingdom and France and used natural uranium metal fuel and magnesium or magnesium alloys for the cladding. Subsequent plants used low-enriched uranium–oxide fuel clad with stainless steel. All 15 of the GCRs operating as of the end of 2014 are located in the United Kingdom. Beginning in the late 1950s in Germany, a new generation of helium-cooled GCRs was developed that used a very robust graphite-coated particle fuel. The coated fuel particles are formed into fuel compacts that are surrounded by a graphite moderator, which typically takes the form of hexagonal blocks (prismatic variant) or spherical balls (pebble-bed variant).

The US constructed and operated a prototype GCR at the Fort St. Vrain plant from 1979 to 1989. It used prismatic block, graphite fuel elements, a uranium–thorium fuel cycle, and helium coolant. Currently the Japanese are operating a 30-MWt prismatic helium-cooled high-temperature test reactor. Three pebble-bed reactors have been constructed and operated: the German AVR and THTR test reactors and the currently operating Chinese HTR-10 test reactor. South Africa built upon the German experience in developing their Pebble Bed Modular Reactor design; unfortunately, financial hardships stalled their efforts to construct a commercial plant. The Chinese, based on their experience with HTR-10, are currently in the process of constructing a pair of 100-MWe pebble-bed reactors in a commercial-sized plant designated the HTR-PM.

Gas-cooled reactors use graphite as a neutron moderator and carbon dioxide gas as the coolant. With the 3% of market share, all of them are installed in the UK. These reactors use natural or slightly enriched uranium as a fuel.

The carbon dioxide circulates through the core, absorbing the heat from the fuel elements and reaching 650 °C. Then, it flows to the heat exchangers, which are located outside of the reactor concrete pressure vessel. These are the gas-to-water heat exchangers that use the once-through principle to boil the water flowing through it. The water is then used for the conventional steam cycle.

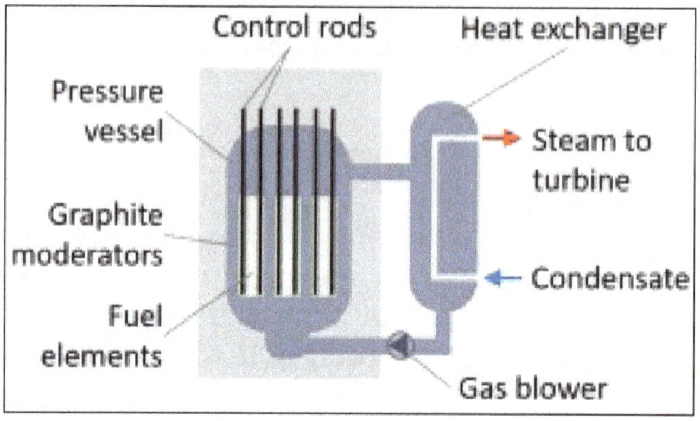

Figure: Schematic diagram of a GCR. Gas-cooled reactor uses gas for inner cycle to bring thermal energy from the fuel elements and transfer it within the heat exchanger to evaporate water.

In this design, boron control rods are used to penetrate the moderator and control the reaction. Besides, there can be a secondary shutdown system involving nitrogen injection into the coolant. In the second generation of the GCR, the steam generators are located inside the concrete pressure vessel, which requires a much bigger structure and therefore higher capital costs.

Gas-cooled reactors have received the greatest attention in the United Kingdom, but also in France and to a lesser extent in the United States, and the USSR. Both natural- and enriched-uranium fuels and CO_2 as coolant and graphite as moderator were used in the U.K. and France. In the United States, enriched fuels and helium coolant are used. Heavy-water-moderated, gas-cooled reactors are also under consideration.

The attractiveness of gas cooling lies in the facts that, in general, gases are safe, are relatively easy to handle, have low macroscopic neutron cross sections, are plentiful and cheap (except in the case of helium), and may be operated at high temperatures without high pressurization.

The main disadvantages are in the lower heat-transfer and heat-transport characteristics of gases, requiring large contact surfaces and flow passages within the reactor and heat exchangers, and their high pumping requirements (between 8 to 20 percent of plant gross power), necessitating careful attention to the problems of fluid flow, pressure drops, etc.

To partially overcome the inherent disadvantages of gas coolants and at the same time to obtain attractive thermodynamic efficiencies, it is necessary to operate the fuel elements at as high temperatures as possible (commensurate with metallurgy) and permit a high gas-temperature rise in the reactor by reducing the gas mass-flow rate and pressurizing the gas. Thus, fuel-element and cladding-material choice and fabrication present major problems, and the trend is toward using ceramic rather than metallic fuels. Also, because gas-cooled reactors are inherently large, they are suited to large-capacity. The reactor itself may impose structural and foundation problems. In general the cycles could be of the Brayton or Rankine types. The greatest majority are of the latter type, using an indirect closed-cycle gas-steam system.

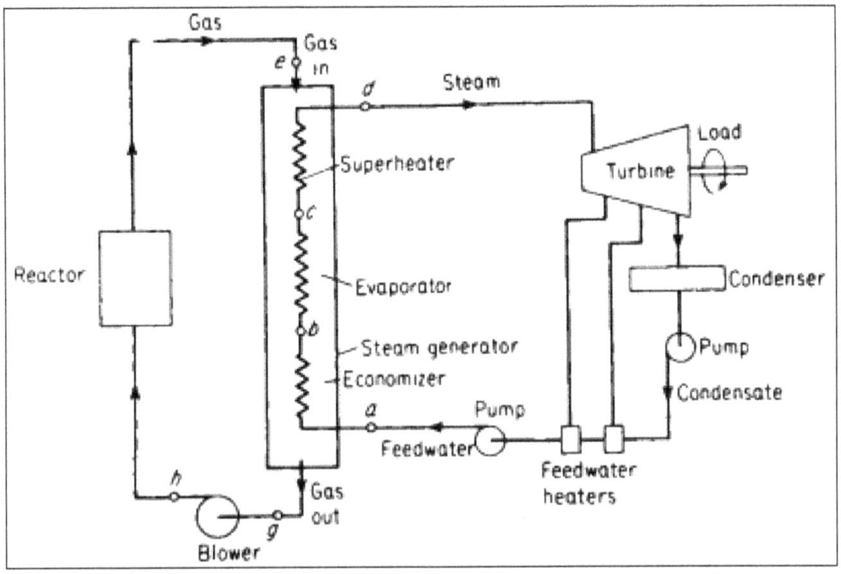

Figure: Schematic of a simple-cycle gassteam-reactor power plant.

The cycle as shown is suited to high-temperature gases. The early British reactors used metallic natural uranium and, because of metallurgical phase changes in the fuel, were suited to low temperature operation. This necessitated the use of a complicated dual-pressure cycle to minimize blower losses and improve cycle efficiency. Such a cycle is shown in figure.

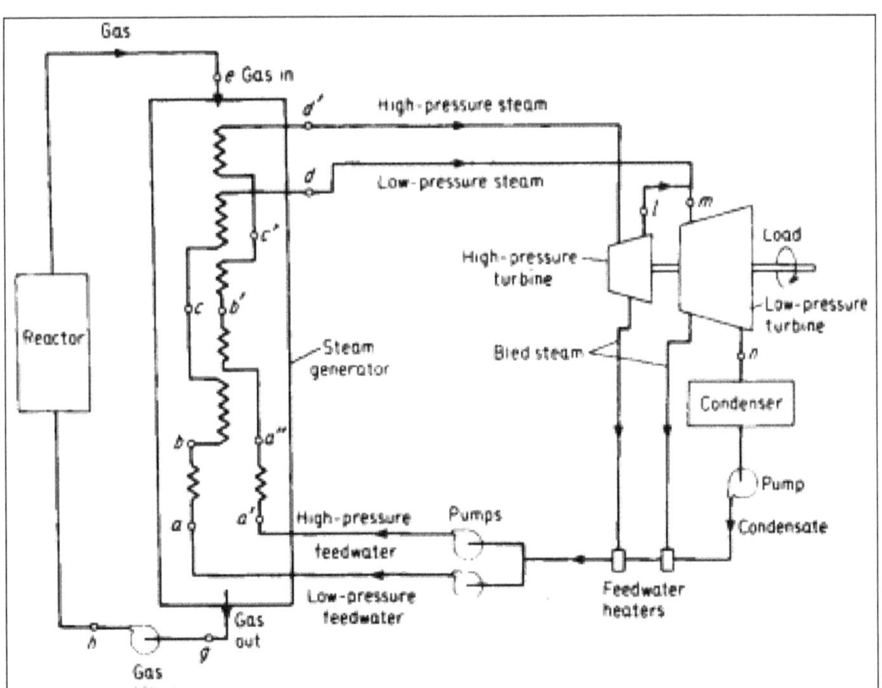

Figure: Schematic of a dual-pressure gas-steam plant.

Magnox Cooled Reactor

Magnox reactors were the first generation of nuclear power plants developed in the UK. They were CO_2-cooled, graphite moderated power reactors with natural uranium metal fuel. The name Magnox refers to the magnesium-aluminum alloy cladding on the metal fuel rods.

The first two Magnox sites, Calder Hall and Chapelcross, each had four dual-purpose reactors primarily intended to produce nuclear material for the UK nuclear weapons program, with the secondary role of generating electricity for UK's national grid. The first unit at Calder Hall went critical in May 1956 and, on 27 August 1956, became the first UK nuclear power plant to connect to the national power grid. All subsequent Magnox plants were larger, two-unit commercial nuclear power plants. The UK's fleet of Magnox plants reached a total of 26 units at 11 sites. On 30 December 2015, the final Magnox plant, Wylfa unit 1, ceased generation after operating for 44 years. This milestone ended 59 years of Magnox reactor operation in the UK.

The only remaining CO2-cooled, graphite moderated commercial power reactors in the world are the UK's Advanced Gas-cooled Reactors (AGRs). Other commercial operators of CO_2-cooled, graphite-moderated reactors have retired their units: Italy in 1987, Spain in 1990, France in 1994, and Japan in 1998. North Korea operates a small CO2-cooled, graphite-moderated reactor that likely has been used for nuclear material production.

A Magnox reactor has a large reactor core that operated at low power density (< 1 kW/liter) and relatively temperatures, which enabled the use of uranium metal fuel and Magnox cladding. The relatively low operating pressure (typically 130 – 150 psig) of the primary circuit enabled the use mild steel for the primary pressure boundary. Here's a comparison of some key parameters for the early Bradwell Magnox reactor and the similar vintage early U.S. pressurized water reactor (PWR), Yankee.

	Bradwell	Yankee
	Magnox	PWR
Initial critically year	1961	1960
Core thermal power (MWt)	538	485
Gross electrical output (MWe)	171	150
Active core diameter (meter/feet)	7.82/25.8	2.3/7.7
Average core power density (kW/liter)	0.58	72
Primary circuit operating pressure (psig)	132	1985

The basic gas and fluid flow paths in the earlier Magnox plants are shown in the following diagram, which shows one of four steam generators. In the closed-loop steel primary circuit, forced circulation of CO_2 transfers heat from the reactor core to the steam generators, which in turn transfer heat into the secondary circuit. In the closed-loop secondary circuit, water delivered to the steam generators is heated and converted to steam, which drives turbine generators to produce electricity. The steam exhausted from the turbines is condensed in the main condensers and returned to the steam generators. An open- or closed-loop circulating water system transfers waste heat from the main condensers to a heat sink (i.e., cooling towers or a body of water).

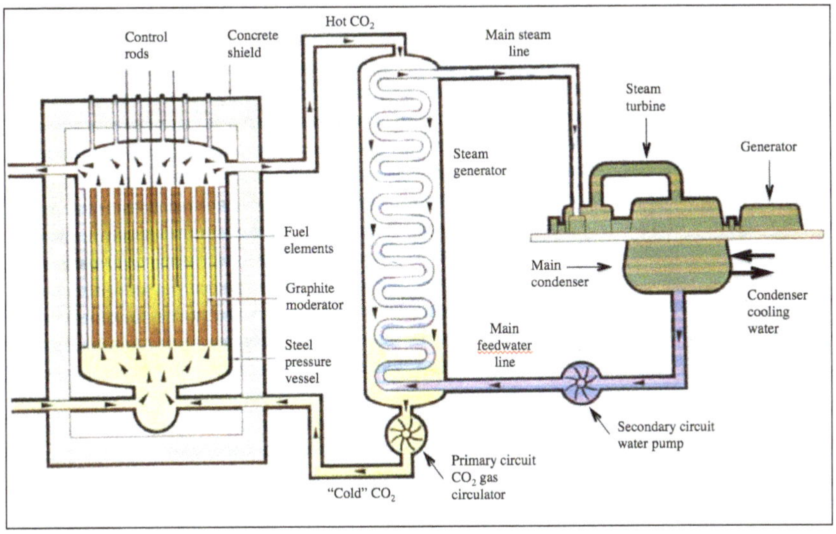

A generic issue for all Magnox plants was the corrosion of mild steel components by the high temperature CO_2 coolant. To manage this issue, the average core outlet gas temperature was reduced from the original design temperature of 414 °C to 360 – 380 °C, with a corresponding decrease in net power output and thermal efficiency.

None of the Magnox reactors are enclosed in a pressure-retaining containment building, as is common practice for most other types of power reactors. In the early Magnox plants, only the reactor was inside an industrial-style building, while the steam generators and parts of the primary circuit piping were outside the building, as shown in the following diagram of a single unit at Calder Hall. The steam generators were enclosed in later plants, primarily to protect them from the weather.

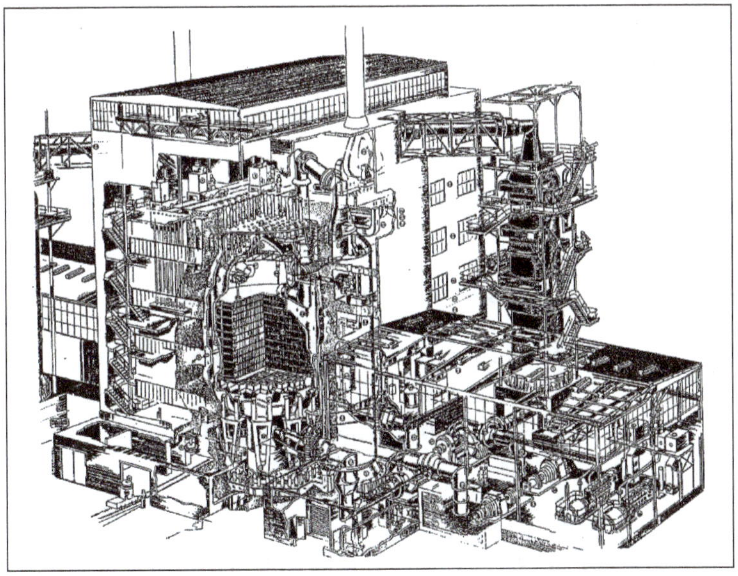

Accident conditions in a Magnox reactor are very different than in a water-cooled reactor. Magnox reactors do not encounter coolant phase change during an accident or have a risk of the core becoming "uncovered" because of a loss of coolant through a breach in the primary circuit. The low core power density, the large heat capacity of the graphite moderator, and the availability of natural circulation flow paths for core cooling limit post-accident core temperatures. On this basis, Magnox reactors were permitted to operate with three barriers to the release of fission products to the atmosphere: The metal fuel matrix, the Magnox fuel cladding, and the mild steel primary circuit pressure boundary.

Export Magnox Plants

The following two single-unit Magnox nuclear power plants were exported to Italy and Japan:

- Latina: Operated from 1963 to 1987; originally rated at 210 MWe, derated to 160 MWe.
- Tōkai 1: Operated from 1966 to 1998; 166 MWe.

UK's Successor to the Magnox

The UK's second generation commercial power reactor is the Advanced Gas-cooled Reactor (AGR), which is a more advanced, higher-temperature, CO_2-cooled, graphite moderated reactor with stainless steel clad, enriched UO_2 fuel. The UK's fleet of AGRs totals 14 units at 6 sites. All are currently operating. Retirement of the oldest units is expected to start in about 2023.

French CO_2-cooled, Graphite Moderated Reactors

In the early 1950s, the UK and France arrived at the same basic fuel/coolant/moderator selection for their first generation power reactors:

- Natural uranium metal fuel.
- CO_2 coolant.
- Graphite moderator.

This choice was driven largely by the desire for nuclear independence and the unavailability of enriched uranium, which the U.S. refused to export, and heavy water (moderator), which was not available in significant quantities.

In France, these reactors were known as UNGG (*Uranium Naturel Graphite Gaz*), which were developed independently of the UK Magnox reactors. All UNGGs used magnesium-zirconium alloy fuel cladding instead of the Magnox magnesium-aluminum alloy.

The first UNGGs were the dual-use Marcoule G2 and G3 reactors, which produced nuclear material for the French nuclear weapons programs and also had a net electric power output of about 27 MWe. The horizontal reactor core was housed in a steel-lined PCRV with external steam generators.

Initial criticality of Marcoule G2 occurred on 21 July 1958, and it first generated electricity in April 1959. Marcoule G3 began operation in 1960. G2 was retired in 1980 and G3 in 1984. Electricité de France (EDF) built six larger UNGG commercial nuclear power plants in three basic configurations:

- The early Chinon A1 and A2 plants had a vertical reactor core in a steel primary circuit.

These plants had net electrical outputs of 70 MWe (A1) and 200 MWe (A2). A1 operated for 10 years from 1963 to 1973. A2 operated longer, from 1965 to 1985.

- The later 480 MWe Chinon A3 plant adopted a different design, with a vertical reactor core in a PCRV with external steam generators. A3 operated for 24 years, from 1966 to 1990.
- The 480 – 515 MWe A1 and A2 plants at Saint Laurent-des-Eaux and the 540 MWe Bugey 1 plant adopted a more advanced and compact design, with an integral primary circuit in a steel-lined PCRV. Unlike the Wylfa Magnox plant, the steam generators were placed under the reactor core in a tall PCRV, as shown in the following diagram. This basic arrangement is similar to the U.S. Fort St. Vrain helium-cooled high-temperature gas-cooled reactor (HTGR) built in the 1970s. Saint Laurent A1 operated from 1969 to 1990, A2 from 1971 to 1994, and Bugey 1 from 1972 to 1994.

Export UNGG

France exported to Spain one UNGG similar to the Saint Laurent-des-Eaux plant. This became the 508 MWe Vandellos unit 1 power plant, which operated from 1972 to 1990.

French Successor to the UNGG

After Bugey 1, France abandoned gas-cooled reactor technology for commercial nuclear power plants. Pressurized water reactor (PWR) technology was chosen for the next generation of French commercial power reactors: the CP0 900 MW PWR.

North Korean CO_2-cooled and Graphite Moderated Reactor

North Korea's Yongbyon nuclear plant is a CO_2-cooled, graphite-moderated reactor with natural uranium fuel. This is a logical choice of reactor type because natural uranium and graphite are domestically available in North Korea. Yongbyon is believed to be a dual-use production reactor that is modeled after the UK's Calder Hall Magnox reactor.

Yongbyon is believed to have a thermal power in the 20 – 25 MWt range and a net electrical output of about 5 MWe. Reactor operation began in 1986. In 2007, operation of Yongbyon was disabled when the cooling towers were demolished to comply with an international agreement related to preventing North Korean production of nuclear material. On 2 April 2013, North Korea announced it would restart Yongbyon. NTI reported that Yongbyon has been operating since September 2013.

Advanced Gas-cooled Reactor

The Advanced Gas Reactor (AGR) represents the generation II of British Gas-Cooled Reactors, developed from the earlier generation I Magnox reactor design of gas-cooled reactors. In the UK there are currently seven AGR nuclear power stations (5 in England and 2 in Scotland) each with two operating reactors. They are all owned and operated by British Energy, and are located and known as Dungeness B, Hartlepool, Heysham 1, Heysham 2, Hinkley Point B, Hunterston B and Torness. These generate nearly 20% of the electricity supplied in the United Kingdom, and the seven power stations of this type are amongst the world's most efficient means of converting the energy of nuclear fission into electricity, as well as being extremely safe.

The first commercial AGR power stations, those at Hinkley Point in Somerset and at Hunterston in Ayrshire began operation in 1976, the latest, at Heysham in Lancashire and Torness in East Lothian in 1988.

Each of the 14 separate reactors is linked with its own 660 MW turbine-generator unit. These turbine-generator units are exactly the same as those to be found at a coal or oil-fired power station of equivalent electrical output (the only difference being the source of heat i.e. at an AGR power station this is nuclear energy and at a conventional power station is the energy of combustion). The end result is the same - the production of heat to generate steam.

Design of the Advanced Gas-cooled Reactor

In a typical AGR system the reactor core, boilers and gas circulators are housed in a single prestressed concrete cavity known as the pressure vessel. The reactor core is a sixteen sided stack of interconnected graphite bricks maintained in position by a steel restraint structure surrounding the core, and is supported by a system of steel plates known as the diagrid. The complete assembly of the core comprises of 332 fuel channels.

The active region of the core containing the fuel consists of 10 years of graphite bricks, each about 825mm high. The active core is enclosed by further graphite which makes up the upper, lower and side neutron shields – the shield is provided in order to safely gain access to the boilers and plant within the pressure vessel when the reactor is shut down and de-pressurized. Additional shielding is provided by steel plates fasted to the steel restraint system. The whole structure is approximately 11m in diameter and 9.8m high and weights around 1300 tonnes.

The graphite core and shielding is surrounded by a steel envelope known as the gas baffle. The diagrid forms the lower end of the gas baffle cylinder. The function of the gas baffle is to provide a route for the coolant gas through the core, and is composed of three main sections: the dome at the top, the cylinder and the skirt. At the dome there are a number of penetrations to allow the passage of fuel assemblies, control rods and to other interstitial channels in the core.

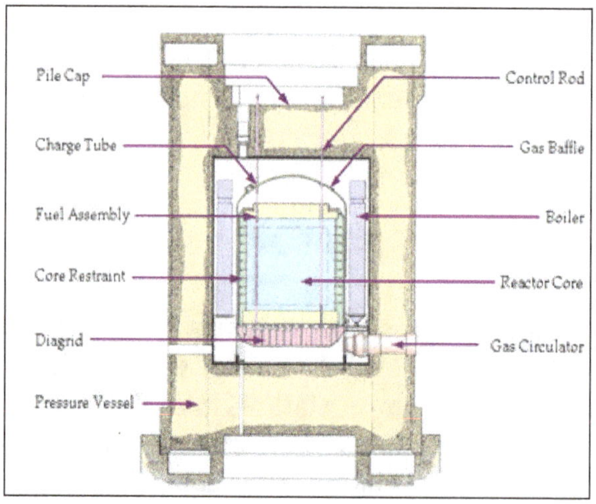

Figure: General schematic of an advanced gas reactor.

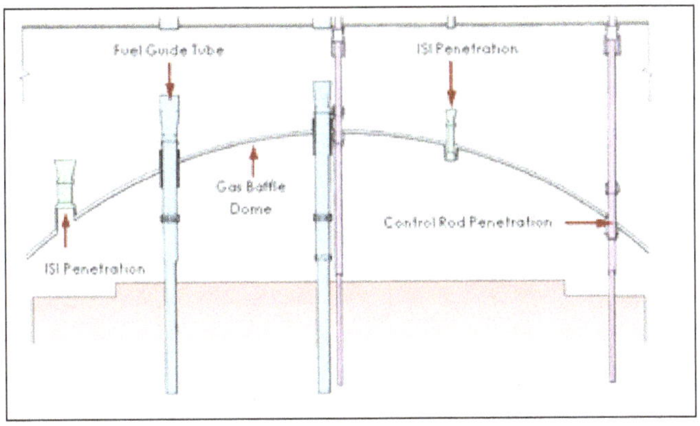

Figure: Penetration at the reactor dome.

Each reactor has 308 fuel channels, 263.5mm in diameter that runs vertically down through the reactor. They are arranged in a regular, square, lattice pattern with a distance of 460mm between their centers. In between the fuel channels there are 81 control rods channels, each 127mm in diameter. Each reactor sits on 109 steel support plates which are in carried in turn by a steel "egg-box" structure – the aforementioned diagrid. The diagrid is supported from pressure vessel floor by 16 diagrid supports.

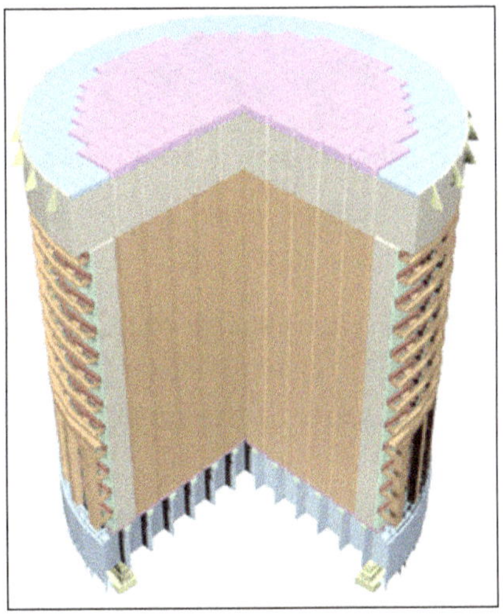
Figure: Virtual buildup of the reactor core showing the peripheral restraint structure.

Figure: Virtual build up of the reactor core supporting diagrid.

Heat is produced within the core of the reactor, and is transferred to pressurized carbon dioxide gas, which is pumped over the fuel by gas circulators around the gas baffle. The gas, in turn, transfers its heat to boilers. Feed water is pumped into the boilers and is converted to superheated steam, which is taken to a conventional 660 MW (e) steam turbine-generator. The energy in the steam is converted to electricity in the turbine-generator, which is supplied to the National Grid.

The Steam

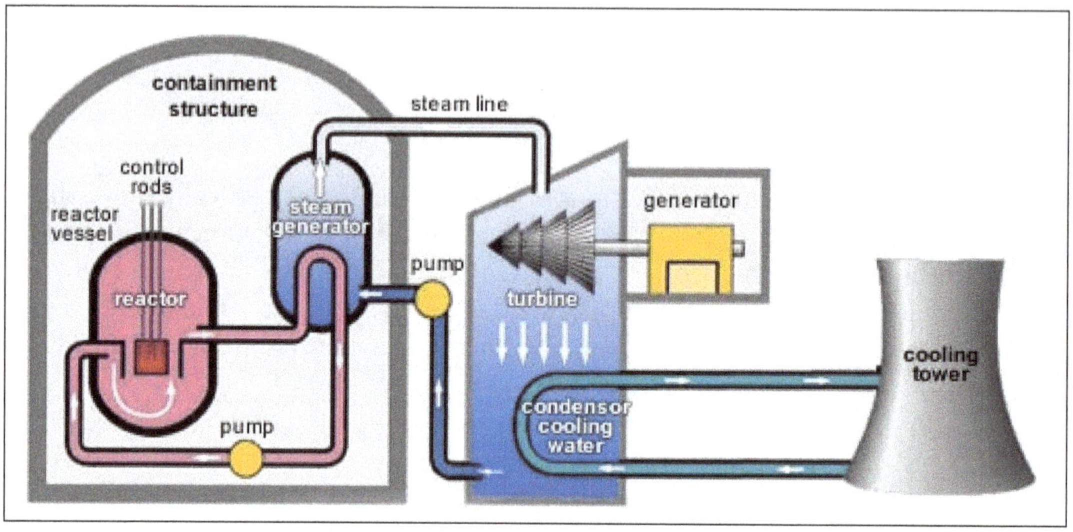

Figure: Nuclear power generation process.

The Reactor Core

The core graphite structure is designed to act as a moderator and provide individual channels for fuel stringer assemblies, control devices and coolant flow for a minimum period of 30 years – these in fact are part of the high level requirements of the reactor core for reactor operation, which are to:

- Provide a structure which, under all operating and fault conditions, ensures that the reactor can be shut down.
- Ensure adequacy of cooling for fuel, control rods, neutron sources, steel absorbers and the graphite must be maintained so that all materials remain within their design limits.
- Maintain unimpeded movement of fuel and control rods to ensure that the above requirements can be met.

Within the core two types of brick are used: one type is a basically circular and contains the fuel channels; interstitial bricks that are basically square and contain control rods, secondary shutdown or coolant holes make up the second type.

The circular bricks are interconnected by loose graphite keys, while the connection between circular and interstitial bricks is through a key that is an integral part of the interstitial brick. Brick shapes and loose keys are optimized for strength and keying is designed to accommodate core movement and seismic loading.

Between each layer of graphite, inter-brick seals are incorporated for the purpose of maintaining the pressure differences generated across the walls of the fuel channels by the appropriate sizing of the flow control ported located in the top and bottom layers of bricks for cooling.

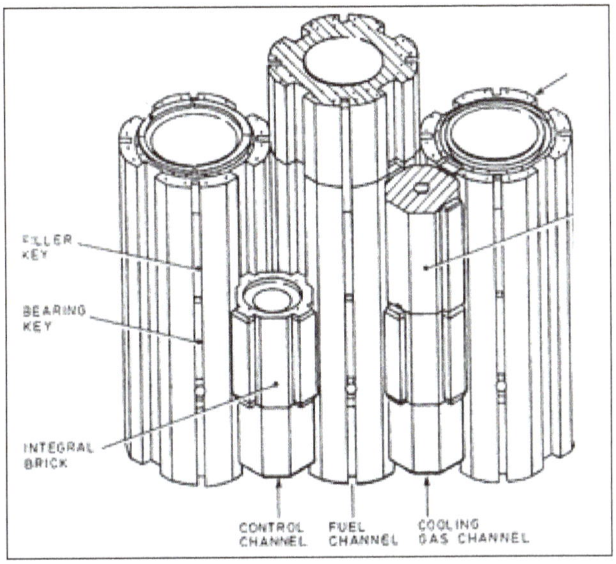

Figure: Diagram of the graphite keying structure.

The overall graphite structure is highly redundant, so that any local failures resulting from, for example, a local high load situation, tight clearances or a faulty component, will not result in gross channel distortions or in the failure of surrounding components. The integrity of the core structure as a whole varies over the life because of the effects of irradiation and radiolytic oxidation.

Core Restraint Structure

The core restraint system is located between the outside of the core and the surrounding boiler shield wall. It maintains the stability of the reactor core. As shown by Figure the lower 3 layers of the core are connected to the vertical restraint columns. These are located at their lower end on keying blocks mounted on the diagrid and at their upper end to a circumferential stiffening ring welded to the boiler shield wall. The remaining layers of the core are connected directly to the circumferential rings welded directly to the boiler shield wall.

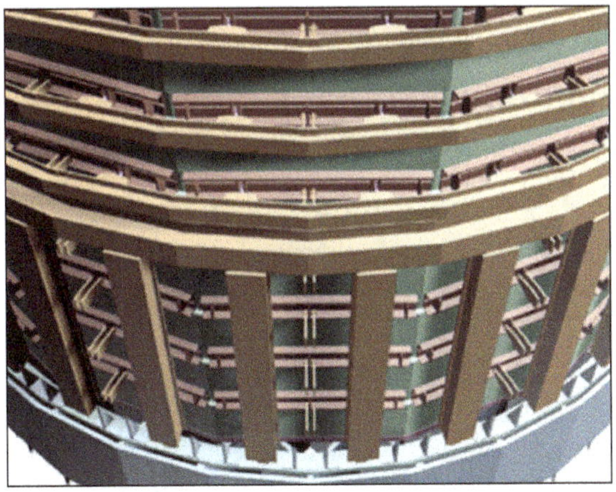

Figure: Virtual build up illustrating the details of the restraint structure at the bottom of the core.

At the core periphery, steel puller rods are located in recesses in selected bricks at each inter-layer position. The rods are in turn attached to restraint beams that form sixteen-sided polygons around the core at each inter-layer position; the beam ends being loosely connected as a secondary restraint feature. Each beam is inset and supported by the graphite and attached to the restraint tank by two spherical ball-ended links (Warwick links), which provide the means of load transfer from the core.

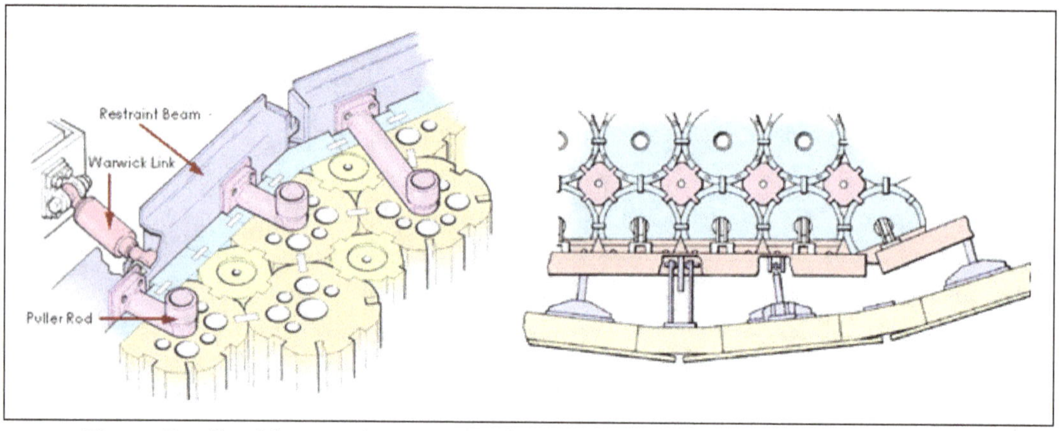

Figure: Details of the restraint structure showing the connection to the graphite bricks.

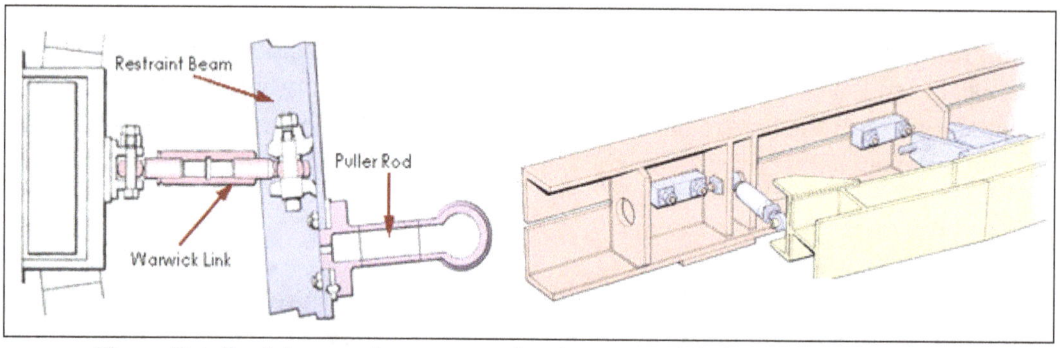

Figure: Details of the restraint structure showing Warwick links and restraint beams.

The restraint structure also exhibits a high degree of redundancy to cater for the possibility of local component failure. These structures do not suffer from the same strict geometrical and material limitations of the core and their integrity can be guarded by the optimum choice and materials and design.

Control and Primary Shutdown Systems

Reactor power is directly proportional to neutron density. In order to control the neutron density neutron absorbing control rods are raised and lowered within the core either by operator or automatic control according to the amount of power required. There are a total of 89 control rods that are housed in standpipes in the top cap of the reactor vessel. These can be divided into the following categories:

- There are a total of 45 "grey control rods" which are weak neutron absorbers and are used as regulating rods i.e. to regulate the power output of the reactor core. These are constantly moving in and out of the core under automatic control to provide a steady power output. 16

of these rods are used as a safety group which can be withdrawn when the reactor is shut down to provide a safeguard against inadvertent criticality.

- "Black control rods" are the primary system for shut down for the reactor core by absorbing a large proportion of thermal neutrons. There are a total of 37 black control rods. This is referred to as the Primary Shutdown Device (PSD).
- There are 7 "sensor rods" that exist as part of the black control rod assembly: these are designed to detect any misalignment in the guide tubes between the reactor core and the steel structures above it when the black rods are being raised or lowered.

The control rod assembly in each standpipe consists of a control rod, a control plug unit, a control rod actuator and the standpipe closure unit. The assembly is designed to be removed with the refueling machine under operating or shut down conditions. The control plug unit is designed to reduce to acceptable levels, radiation streaming from within the core through the standpipe penetrations in the vessel roof. It consists of steel plug with a central hole through which passes the control rod suspension chain.

Each actuator is complete with motor-operated winding gear and suspension chain storage, electromagnetic clutch, hand-winding drive to the clutch, rod position indicator and limit switches. The actuator and rod drive is design frequent small movements, and the speed of the control rod is controlled by the induction motor. In the event of a reactor trip, the clutch is also tripped to allow control rod insertion by gravity, which is controlled by a carbon disc brake and centrifugal mechanism.

Secondary Shutdown Systems

As a backup measures against failure of the PSD there exist several Secondary Shutdown (SSD) devices. Different systems have been installed in different reactor cores, the two most common being the following:

The nitrogen injection system consists of an arrangement of pipework and control values linking a permanent nitrogen gas store to a group of 165 interstitial channels in the core. The store, consisting of banks of high pressure cylinders, is common to both reactors and holds sufficient nitrogen in gaseous form for the shutdown and subsequent hold down of one reactor, provided it remains pressurized. Approximately half an hour is required for full deployment of this system, which is completed in two stages.

The first stage is on trip where the nitrogen flows through feed pipes to each of the interstitial secondary shutdown channels. When the first trip values are opened the initial flow purges these channels of carbon dioxide and fills them with nitrogen. The second stage provides additional flow to each channel as the nitrogen flow from within the channels is forced into the re-entrant passages and through the fuel channels, thus gradually building up the nitrogen concentration in the coolant gas circuit until it is sufficient to hold the shutdown core in a sub-critical condition for several hours.

The boron bead device consists of a store loaded with boron glass beads of 3mm diameter which are injected into 32 of the secondary shutdown channels in the reactor core. Each of these channels has an associated delivery pipe, with one end terminated at the top of the channel and the other end connected to one of the bead storage hoppers. The hoppers are located below the reactor pressure vessel in the secondary shutdown room.

On trip, the carbon dioxide is used to convey the beads pneumatically from the bottom of the hopper to the top of the channel. The beads are directed downloads into the channel from the open end of the delivery pipe until the channel is filled. Sufficient gas is provided by the station coolant system to inject all the stores of beads simultaneously. For their recovery, the beads are passed from the bottom of each channel through a vertical recovery pipe terminating with an isolating ball valve in the secondary shutdown room.

Post Trip Cooling

As with all nuclear power stations after shutdown and fault-initiated trips the fission products still undergo radioactive decay and continue to release heat (decay heat) even after fission. Post-trip heat removal systems are provided to removal the residual heat so as to prevent the fuel overheating and damage to the reactor structures. The main functions of these systems are to firstly circulate sufficient gas coolant to transfer the heat from the fuel and elsewhere to the boilers and secondly to provide sufficient feed water to the boilers to enable this heat to be transferred to the environment. Note that AGRs are capable of using natural circulation to prevent overheating, provided the main boilers have feed water available.

There are two main systems for post trip cooling: these systems are arbitrarily designated 'X' and 'Y' systems. The 'X' system is the preferred system for cooling when the reactor is pressurized. Heat is removed from the fuel by forced circulation provided by the gas circulators operating at low speeds. The heat is removed from the reactor coolant by the decay heat boilers, or which there are four, situated directly below and separate from the main boilers in the boiler annulus.

The 'Y' system is used if either de-pressurization of the reactor occurs or if the 'X' system fails. Heat is removed by forced circulation provided by the gas circulators operating a high speeds to compensate for the lack of pressure. The heat is removed from the reactor coolant by the main boilers fed by the emergency boiler feed system.

Fuel Assemblies

The slightly enriched (2 – 3%) uranium used as fuel in AGRs is made up in the form of cylindrical pellets. These are contained within sealed, stainless-steel tubes to form fuel pins. Each pin is approximately 980 mm in length. 36 fuel pins are arranged in clusters within graphite sleeves to form each fuel element, eight of which are linked together with a tie bar to form a fuel stringer.

At the top of the fuel stringer is the fuel plug unit. This comprises of a closure unit, a biological shield plug to limit neutron and gamma radiation through the standpipe, a gag unit to adjust the coolant flow through the individual channels, and neutron scatter plug to prevent neutrons streaming up the fuel channel. Gas-sampling instrumentation and channel gas outlet thermocouples are also incorporated into the unit to provide monitoring.

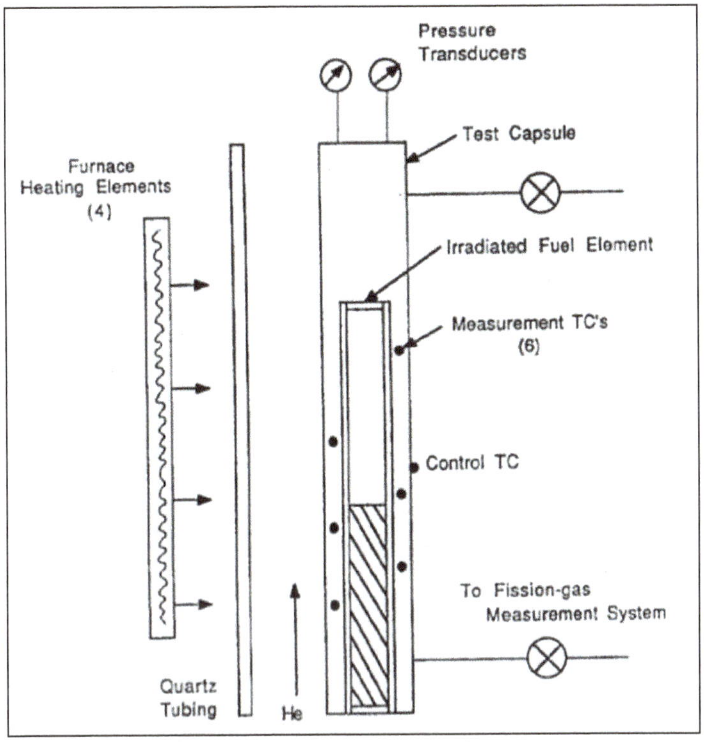

Figure: Schematic of a fuel pin.

Figure: Photographs of the fuel pin.

Insertion of the fuel assembly into the reactor core is handled by a single re-fuelling or charge machine. The machine runs on a traveling gantry that spans the width of the charge hall and is supported on rails running parallel along the length of the hall. The charge machine is designed to handle both fuel and control assemblies and is capable of refueling 120 channels per year.

During normal operation, the fuel plug closure units seal the reactor pressure vessel boundary at each fuel standpipe. A primary lock is provided on the refueling machine grab, and a secondary lock is provided by a retractable drive mechanism built into the standpipe extension tube.

Figure: Photograph of the charge machine.

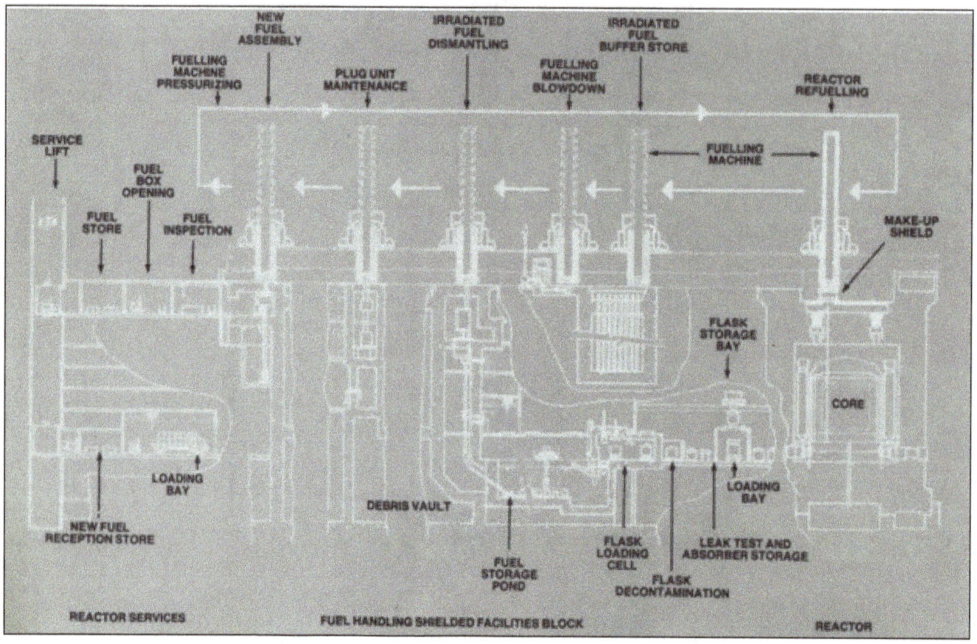

Figure: Torness/Heysham B fuel flow cycle.

Primary Heat Cycle (Gas Sided)

Carbon dioxide is the gas used for transferring the heat produced form the reactor core to the boilers. The gas is pumped through the gas baffle channels of the reactor core at high pressure by 8 gas circulators. There are two circulators per each quandrant. These are housed in large penetrations towards the bottom of the pressure vessel, and each one forces the CO_2 around the primary circuit at a rate of around 500kg every second.

Cooled gas is drawn from the bottom of the boilers by the gas circulators and is discharged into the space below the core. About half of this has flows directly to the fuel channel inlets, while the remainder, known as the re-entrant flow, passes up the annulus surrounding the core and returns downwards through the core in the passages between the graphite bricks to rejoin the main coolant flow at the bottom of the fuel channels. The re-entrant flow is thus to cool the graphite bricks, the core steel restraint system and the gas baffle.

The combined flow passes up the fuel channels and is led through the space between the top of the core and the gas baffle in guide tubes that discharge the hot gas into the area above the gas baffle and thence to the top of the boilers.

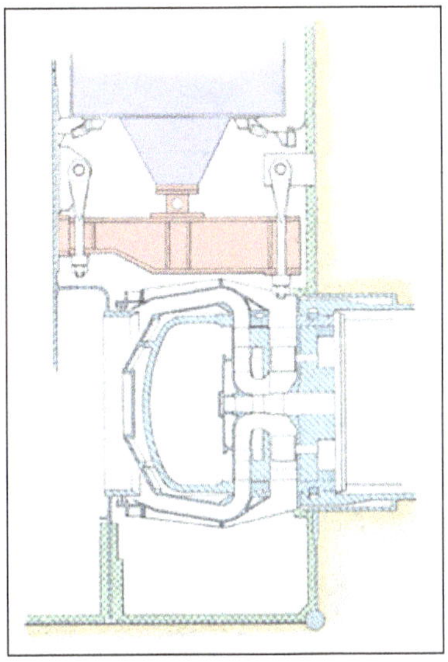

Figure: Gas circulator.

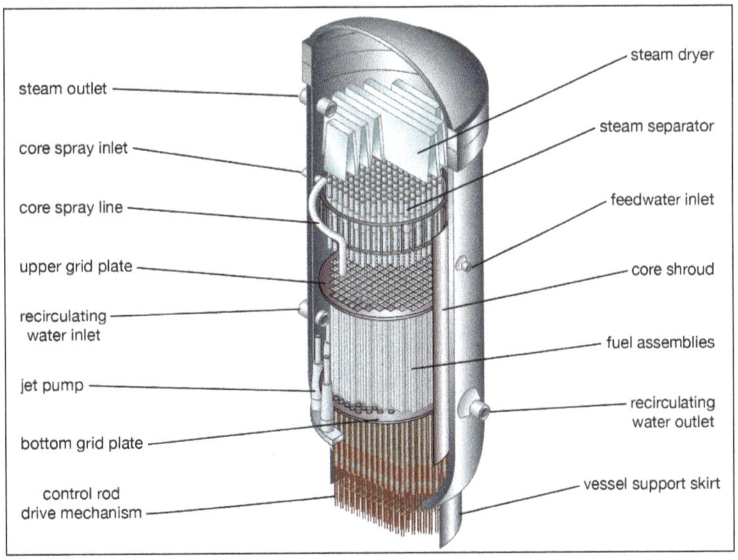

Figure: Primary heat cycle through the reactor core.

Secondary Heat Cycle (Steam Side)

Each reactor has 12 boilers situated between the gas baffle and the pressure vessel liner and is partitioned into four quadrants, each containing one boiler and two gas circulators. Within each quadrant there are three main boiler units, each consisting of an economizer, evaporator and superheater sections. The boilers are of the once-through type.

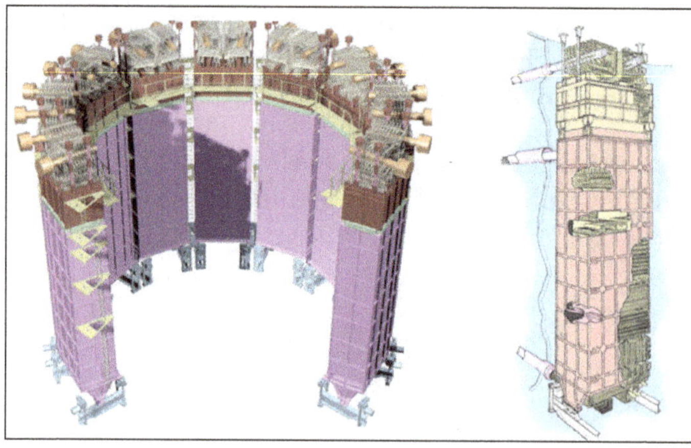

Figure: Boilers around the periphery of the reactor core.

Each high pressure unit is supported by bearings mounted on a pair of boiler support beams, and the reheater units are suspended from the roof by four support slings.

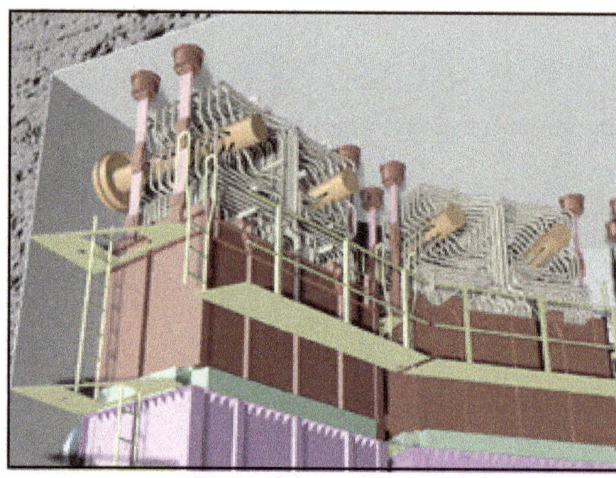

Figure: Reheator hanger.

Figure: Boiler support beam.

The hot carbon dioxide from the reactor flows into the boilers at around 4 tonnes per second and at a temperature of approximately 600 °C. When the CO_2 exits the boiler it is some 300 °C cooler – this generators half a tonne of steam every second.

Pressure Vessel

The design of a gas-cooled reactor calls for pressure vessels to withstand the pressure of circulating coolant gas used to transport heat from the reactor core to the main boilers and thick concrete shielding to absorb neutrons and gamma radiation given off during fission. These requirements can be combined and in the AGRs, prestressed concrete pressure vessels are used to carry out these functions.

The pressure vessel walls are approximately 5.8m thick, the top slab 5.4m, and the bottom slab 7.5m. The prestressing and post-tensioning system consists of around 3,600 steel tendons in helical formation threaded through 76mm mild-steel tubes that are embedded in the concrete during construction. The tendons are anchored in stressing galleries at the top and bottom of the vessel, which provide access to the tendons and from which insertion and stressing can take place. Routine checking of the tendons is carried out throughout the life of the reactor. All the tendon strand anchorage loads can be checked individually and tendons re-tensioned or replaced if need be. The number of tendons is very much in excess of those necessary to provide the requisite strength, so that in theory many could fail without fear of pressure vessel failure.

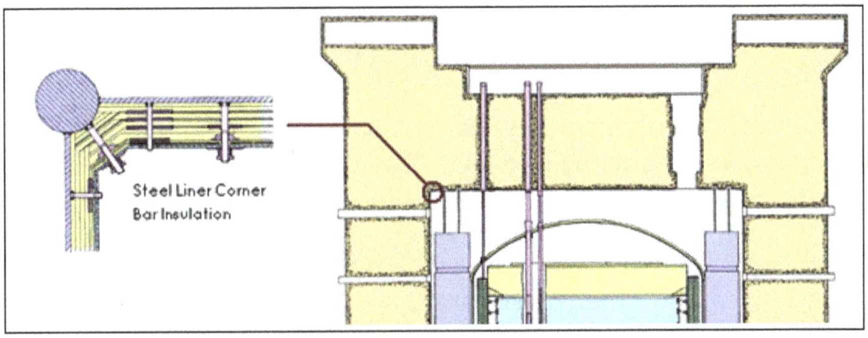

Figure: Corner liner details inside the pressure vessel.

On the inside of the vessel there is a steel liner that is gas-tight – the main purpose of which is to provide a leak-tight membrane to prevent the release of hot carbon dioxide gas through the concrete and penetrations, therefore minimizing the risk of release of radioactivity from the plant, as well as serving as a foundation for the cooling and insulation systems that protect the concrete from excessive temperatures and temperature gradients.

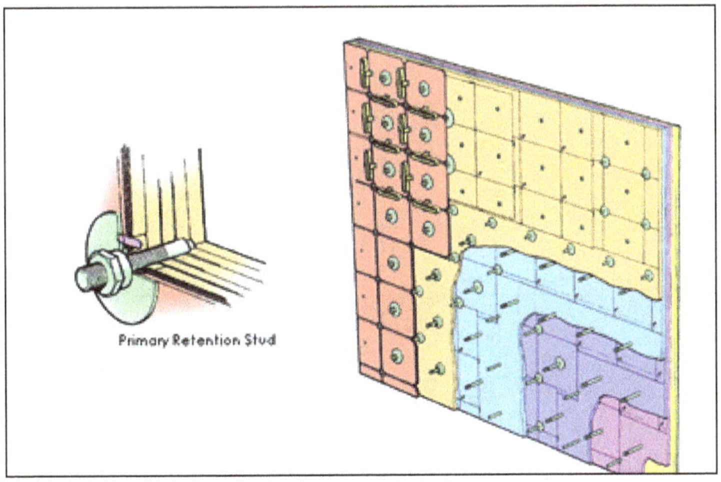

Figure: Layers of the liner.

The liner is insulated on the inner face and water cooled on the outside. The insulation consists of layers of ceramic fibers held in contact with the liner by a system of steel foils and cover plates, each of which is retained in position by a series of primary and secondary retention systems.

The cooling system is a closed-loop system circulated treated water through a network of pipes welded to the vault liner plate so as to provide full-length heat transfer paths from the metal surface to the circulating water. The water circulating through the network of pipes removes the heat passing through the insulation attached to the inner surface of the liners. Together, the insulation and liner cooling system ensure that the liners and vessel concrete are maintained at acceptable temperatures.

High-temperature Gas-cooled Reactor

The high-temperature gas-cooled reactor (HTGR) is similar in concept to the AGR. It uses uranium fuel, a graphite moderator and a gas as a coolant. In this case, however, the gas is helium. Several attempts have been made to build reactors of this type, but none has so far entered commercial service. Early development work was carried out in the United States. The US design utilised fuel elements in the shape of interlocking hexagonal prisms of graphite containing the fissile material. HTGR fuel is often much more highly enriched than the fuel in a water-cooled reactor, with up to 8% uranium-235. The arrays of hexagonal graphite prisms contain shafts for control rods and passages for the helium to pass through and carry away the heat generated by fission.

Another design, developed in Germany, uses uranium oxide fuel which is sealed inside a graphite shell to form a billiard ball-sized fuel element called a pebble. This gives the reactor its name, the pebble-bed reactor. A schematic of a reactor of this type is shown in Figure. Development of this in Germany was eventually abandoned, but the idea was taken up during the 1990s by the South African utility Eskom which continued development until 2010 when it appeared to have been halted. Japan and China have funded experimental programmes too.

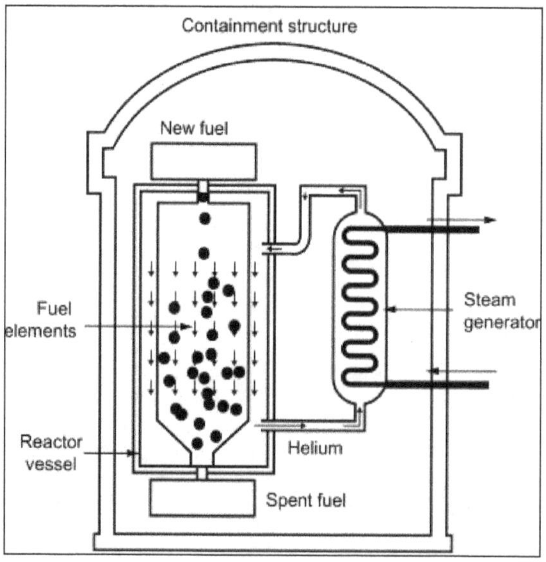

Figure: Schematic of an HTGR.

The advantage of the HTGR is that the moderator, graphite and the coolant, helium, can operate at high-temperature without reacting or deteriorating. A typical HTGR will operate at a pressure of 100 atmospheres and at a temperature of up to 900 °C. This enables better thermodynamic conditions to be achieved, leading to higher efficiency. The reactor is designed so that in the event of a coolant failure it will be able to withstand the rise in internal temperature without failing.

The HTGR can use a dual cooling cycle system in which the helium coolant passes through a heat exchanger where the heat is transferred to water and steam is generated to drive a steam turbine. This arrangement is around 38% efficient. However, a more advanced system uses the helium directly to drive a gas turbine before using the remaining heat to raise steam. This arrangement is sometimes called a gas turbine modular helium reactor (GT-MHR). In theory the GT-MHR can achieve an energy conversion efficiency of 48%.

One of the advantages of the HTGR is that it can be built in relatively small unit sizes. Modules can have generating capacities of between 100 and 200 MW, making them attractive for a wider variety of applications. The modular form of most designs also makes it easy to expand a plant by adding new modules. However, no reactors of this design have yet entered commercial service.

Very-high-temperature Reactor

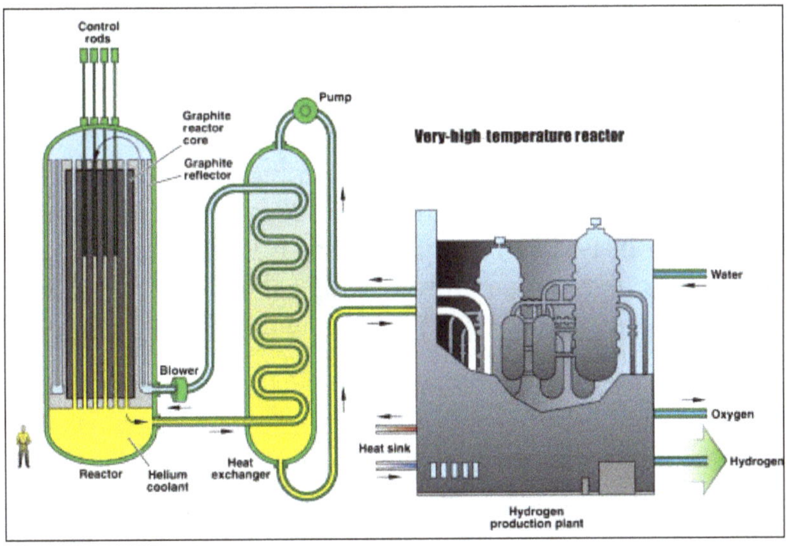

Figure: Very-high-temperature reactor (VHTR), Helium-gas-cooled, graphite-moderated, thermal-neutron-spectrum reactor with core outlet temperature 900–1000 °C (shown with hydrogen cogeneration).

The very-high-temperature reactor (VHTR) is a further step in the evolutionary development of high-temperature reactors (HTRs). The VHTR is a helium-gas-cooled, graphite-moderated, thermal-neutron-spectrum reactor with a core outlet temperature > 900 °C, and a goal of 1000 °C, sufficient to support high-temperature processes such as production of hydrogen through thermochemical processes. The thermal power of the reactor is set at a level that allows passive decay-heat removal, currently estimated to be about 600 MWth. The VHTR is useful for cogeneration of electricity and hydrogen, as well as for other process-heat applications, related to the chemical, oil, and iron industries. It is able to produce hydrogen from water by using thermochemical, electrochemical, or hybrid processes with reduced emission of CO_2 gases. At first, a once-through low-enriched uranium (LEU) (< 20% U235) fuel cycle will be adopted, but a closed fuel cycle will be assessed, as well as potential symbiotic fuel cycles with other types of reactors (especially, light-water reactors (LWRs)) for waste-reduction purposes. The system is expected to be available for commercial deployment by 2020.

The technical basis for VHTR is the tri-isotropic (TRISO)-coated particle fuel. The VHTR has potential for inherent safety, high thermal efficiency, process-heat-application capability, low operation and maintenance costs, and modular construction.

In general, the reactor core of the VHTR can be a prismatic block core such as the Japanese high-temperature test reactor (HTTR), or a pebble-bed core such as the Chinese HTR-10. For electricity generation, a helium-gas-turbine system can be directly set in the primary-coolant loop, which is called a direct cycle, or at the lower end of the outlet-temperature range, a steam generator can be used with a conventional Rankine cycle. For nuclear heat applications such as process heat for refineries, petrochemistry, metallurgy, and hydrogen production, the heat application process is generally coupled with the reactor through an intermediate heat exchanger (IHX), the so-called indirect cycle. The VHTR can produce hydrogen from only heat and water by using thermochemical processes (such as the Sulfur-Iodine (S-I) process or the hybrid-sulfur process), High-temperature steam electrolysis (HTSE), or from heat, water, and natural gas by applying the steam-reformer technology.

While the original approach for VHTR at the start of the Generation IV program focused on very high outlet temperatures and hydrogen production, current market assessments have indicated that electricity production and industrial processes based on high-temperature steam that require modest outlet temperatures (700–850°C) have the greatest potential for application in the next decade, and, also, reduce technical risk associated with higher outlet temperatures. As a result, over the past decade, the focus has moved from higher outlet temperature designs such as gas-turbine modular helium reactor (GT-MHR) and pebble-bed modular reactor (PBMR) to lower outlet temperature designs such as high-temperature reactor pebble-bed modules (HTR-PM) in China and the Next-Generation Nuclear Plant (NGNP) in the United States.

The VHTR has two typical reactor configurations, namely: (1) the pebble-bed type and (2) the prismatic-block type. Although the shape of a fuel element for two configurations is different, the technical basis for both configurations is same, such as the TRISO-coated particle fuel in the graphite matrix, full ceramic (graphite) core structure, helium coolant, and low power density, in order to achieve high outlet temperature and the retention of fission production inside the coated particle at normal operation condition and accident condition. The VHTR can support alternative fuel cycles such as U-Pu, Pu, MOX, and U-Th.

Gas-cooled Fast Reactor

The GFR system is a high-temperature helium-cooled fast-spectrum reactor with a closed fuel cycle. It combines the advantages of fast-spectrum systems for long-term sustainability of uranium resources and waste minimisation (through fuel multiple reprocessing and fission of long-lived actinides), with those of high-temperature systems (high thermal cycle efficiency and industrial use of the generated heat, for hydrogen production for example).

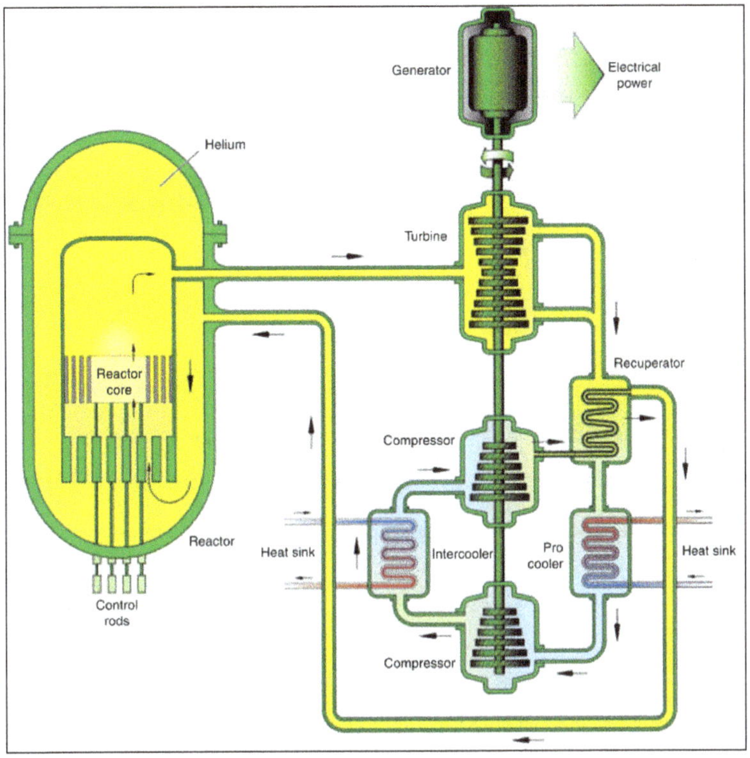

The GFR uses the same fuel recycling processes as the SFR and the same reactor technology as the VHTR. Therefore, its development approach is to rely, in so far as feasible, on technologies developed for the VHTR for structures, materials, components and power conversion system. Nevertheless, it calls for specific R&D beyond the current and foreseen work on the VHTR system, mainly on core design and safety approach.

The reference design for GFR is based around a 2 400 MWth reactor core contained within a steel pressure vessel. The core consists of an assembly of hexagonal fuel elements, each consisting of ceramic-clad, mixed-carbide-fuelled pins contained within a ceramic hex-tube. The favoured material at the moment for the pin clad and hex-tubes is silicon carbide fibre reinforced silicon carbide. The figure below shows the reactor core located within its fabricated steel pressure vessel surrounded by main heat exchangers and decay heat removal loops. The whole of the primary circuit is contained within a secondary pressure boundary, the guard containment.

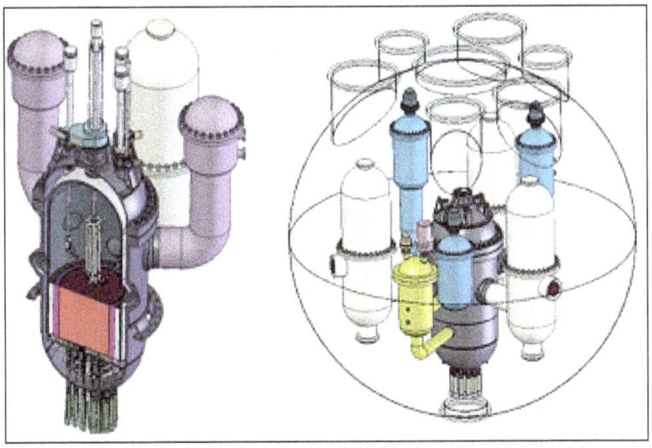

The coolant is helium and the core outlet temperature will be of the order of 850 °C. A heat exchanger transfers the heat from the primary helium coolant to a secondary gas cycle containing a helium-nitrogen mixture which, in turn drives a closed cycle gas turbine. The waste heat from the gas turbine exhaust is used to raise steam in a steam generator which is then used to drive a steam turbine. Such a combined cycle is common practice in natural gas-fired power plant so represents an established technology, with the only difference in the GFR case being the use of a closed cycle gas-turbine.

References

- Gas-cooled-reactors, engineering: sciencedirect.com, Retrieved 12, April 2020
- Guide-to-Graphite, graphiteknowledgebase: iaea.org, Retrieved 21, June 2020
- High-temperature-gas-reactors, engineering: sciencedirect.com, Retrieved 04, January 2020
- Gas-cooled-fast-reactor-42148: gen-4.org, Retrieved 10, July 2020

Chapter 4

Breeder Reactors

A breeder reactor is a nuclear reactor which produces more fissile materials than it consumes to generate energy. They are designed to extend the nuclear fuel supply for the generation of electricity. Fast breeder reactors use fast neutrons to generate nuclear fuels and power. In thermal breeder reactors the fission chain is sustained by thermal electrons. The topics elaborated in this chapter will help in gaining a better perspective about breeder reactors and their types.

Breeder reactors are a type of nuclear reactor which produces more fissile materials than they consume. They are designed to extend the nuclear fuel supply for the generation of electricity, and have even been mistakenly called a potential renewable energy source. Breeder reactors certainly have the ability to make nuclear fuels quite sustainable. Concerns about nuclear weapons proliferation have been one large impediment to creating commercial breeder reactors.

Unlike normal reactors which only use uranium-235 as their fuel, which is only available in scarce concentrations of around 0.7% of natural uranium without enrichment, breeder reactors also make use of natural uranium-238 which is much more common. They can use approximately 70% of the uranium-238 for production of power, whereas normal reactors can only use around 1% of it. They can also use thorium-232 to breed uranium-233, another fissionable product. The most common breeding is of plutonium-239, which is bred through the process seen in Figure below.

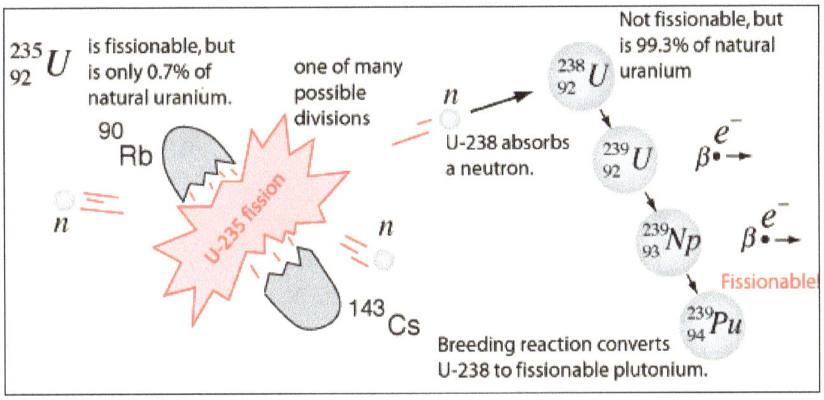

Figure: The breeding of 239Pu in a breeder reactor. The neutrons are supplied by the decay of 235U, which transmute 238U to plutonium.

The first experimental breeder reactor (EBR-1) developed was in 1951 in Idaho, U.S.A. Subsequently Russia, Japan, Great Britain and France all developed experimental breeder reactors; however no nation has developed one suitable for high-capacity commercial use. So far, France has made the largest implementation of breeder reactors with their Super-Phenix fast breeder reactor.

Types

There are two categories of breeder reactors, based on the speed of the neutrons. Fast breeder reactors which use uranium-238 as fuel and thermal breeder reactors which use thorium-232 as fuel. Fast breeders do not require moderation since the neutrons need to be moving fast, whereas thermal breeders make us of moderation to achieve slower-moving neutrons.

Fast

The most promising type of breeder reactor is the Liquid Metal Fast Breeder Reactor (LMFBR), which operates by using liquid sodium as its coolant, and breeds plutonium from uranium-238. It works by using highly enriched uranium, between 15-20% uranium-235 content, surrounded or "blanketed" by natural uranium-238 in the reactor core. No moderator is used to slow down the neutrons, because fast neutrons transmute uranium-238 much more efficiently than slow neutrons.

Using water as a coolant would reduce the neutron abundance, since neutrons are absorbed by water. Therefore liquid sodium is used instead. This immediately raised concerns of safety when initially thought of, since sodium is a highly reactive element. It is important to keep the liquid sodium from contact with air or oxygen to avoid explosions; however they aren't any more dangerous than pressurized water reactors. This is because the sodium doesn't need to be pressurized to remain in a liquid state like water does; its boiling point is 892 °C. This makes the bursting of pipes far less likely than in other water-reactors. Liquid sodium is also a very good choice because of its heat transfer capabilities, due to its high specific heat capacity. Other fast breeder types include supercritical water cooled reactors, molten salt reactors, and gas-cooled reactors.

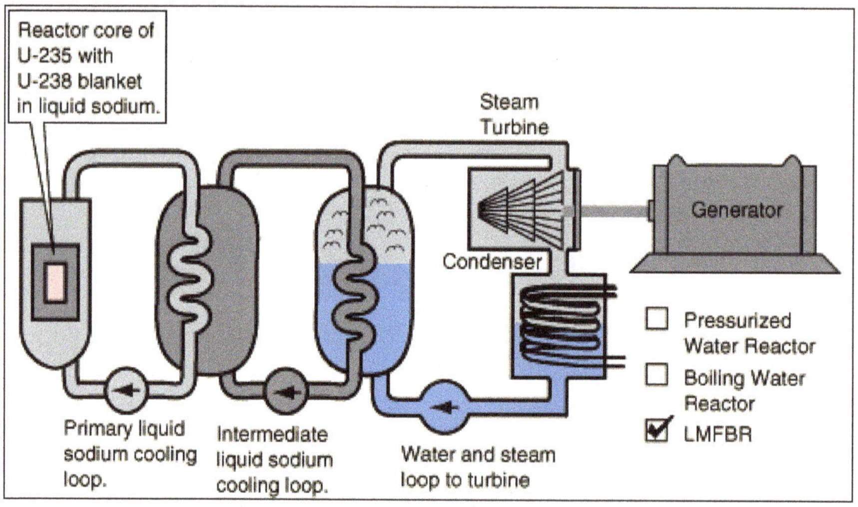

Figure: Liquid metal fast breeder reactor.

Thermal

The possibility to breed fissile material in slow neutron reactors is unique to thorium, as uranium cannot use thermal neutrons to do so. These reactors would need a fissile material to start the breeding just like fast breeder reactors, however the neutrons produced from this fissile material would need to be slowed down by a moderator.

The technology is much simpler than that of the liquid metal fast breeder; light water is used as the coolant to remove the heat produced by the continuous series of fission reactions rather than a liquid metal system.

Thorium hasn't been used in large scale reactors; however some reactors have used it successfully in the past. A light water breeder reactor in Shippingport, Pa. USA operated for 5 years, and by the end of its operation it had 1.4% more fissile fuel than it began with.

Breeding Ratio

An important concept for a breeder reactor is how much fissionable fuel is being produced compared to how much fuel is being used. This is known as the breeding ratio. For example for the breeding of plutonium, the ratio would be the amount of plutonium produced to the amount of uranium-235 used. In the liquid metal fast breeder reactor (LMFBR), the breeding ratio is 1.4; however the actual achieved ratio is around 1.2.

The number 1.4 is based off of the average number of neutrons given off by a fission reaction of uranium-235, which is 2.4. Only 1 neutron is needed for the fission chain reaction to be stable, so the remaining 1.4 neutrons (on average) could be used for the breeding of uranium-238.

Doubling Time

The amount of time for a breeder to produce enough material to fuel a second reactor is called its doubling time. Present goals for a breeder reactor's doubling time is 10 years, which means it would operate for 10 years producing energy and breeding fissionable material, after which this material could be used to produce the same energy output for another 10 years.

Fast Breeder Reactor

Reactors known as fast neutrons or breeders reactors do not need to slow down neutrons. With fuel enriched to 15 or 20% in uranium-235 or plutonium, it become no longer necessary to "thermalize" neutrons as in the common pressurized water reactors, for example. The major interest of fast neutrons is that they are able to extract all the fission energy contained in uranium of the earth's crust.

Fast neutron reactors use mainly plutonium as nuclear fuel. Since they do not require any moderator, their core is 10 times more compact than those of conventional civilian reactors. Cooling is achieved with sodium or with molten lead. This cooling poses a safety problem, particularly concerning sodium, which reacts violently with air and water. But a breeder like PHENIX has been operating 40 years in France without accident, before its final shut down in 2011.

The french breeder reactor SUPERPHENIX at Creys-Malville SUPERPHENIX was a major breeder reactor designed to produce electricity. In 1996, it was decided to abandon the production of electricity, but to use fast neutrons from the reactor to burn plutonium and radioactive waste. This option was abandoned in 1997.

Their main interest is to generate under certain conditions more fissile material than consumed. Such reactors are called "breeders". Their principle is particularly attractive for power production, since breeders offer a virtually inexhaustible source of energy: one of the oldest fantasies of mankind, like a motor running with water. It is estimated that one could extract with them 50 to 80 times more energy than with today ordinary reactors using slow neutrons to burn uranium-235.

Fast neutrons are also effective to burn the long-lived minor actinides of radioactive waste. With a breeder, the amount of plutonium remains in principle constant, because the fuel is regenerated. But regeneration is not mandatory. If regeneration is abandoned, it becomes also possible to burn plutonium partially or completely.

The first fast neutron reactor that supplied electricity at an industrial scale was the French breeder Superphenix. It was stopped in 1997 more for political than technical reasons. Various countries, including France, Japan and Russia are continuing research on this sector in the context of studies of fourth generation reactors. Russia operates in the Urals at Beloyarsk two fast reactors, BN-600 in operation since 1980 and BN-800 a powerful 883 megawatt unit commissioned in 2016.

Beloyarsk Nuclear Power Plant: BN-600 Main building of the Russian Beloyarsk nuclear power plant near Sverdlovsk in the Russian Urals. The plant has a fast neutron reactor (BV-600) since 1980. A second more powerful unit ((BN-800) of 883 megawatt is in operation since 2016 and is now the most powerful operating in the world. A third one is under consideration.

Fast neutron reactors could be started, when the time comes, using existing stocks of depleted uranium by-products of the isotopic separation process and of plutonium separated in the reprocessing of spent nuclear fuel. The amount of depleted uranium and of the uranium coming from the reprocessing, would be enough to make run for several thousands of years without using natural uranium. The plutonium of spent MOX fuel, currently in storage, could also be blended to 15% with depleted uranium.

As waste burners, fast neutron reactors may have to compete with hybrid reactors, whose principle was proposed by the Nobel Prize Carlo Rubbia in 1993. These hybrid reactors might be dedicated to the destruction of actinides. They would have the ability to destroy plutonium and would additionally be safe.

Difference between Fast Breeder Reactors and Regular Nuclear Power Plants

Nuclear reactors generate energy through fission, the process by which an atomic nucleus splits into two or more smaller nuclei. During fission, a small amount of mass is converted into energy, which can be used to power a generator to create electricity. In order to harness this energy, a controlled chain reaction is required for fission to take place. When a uranium nucleus in a reactor splits, it produces two or more neutrons that can then be absorbed by other nuclei, causing them to undergo fission as well. More neutrons are released in turn and continuous fission is achieved.

Neutrons produced by fission have high energies and move extremely quickly. These so-called fast neutrons do not cause fission as efficiently as slower-moving ones so they are slowed down in most reactors by the process of moderation. A liquid or gas moderator, commonly water or helium, cools the neutrons to optimum energies for causing fission. These slower neutrons are also called thermal neutrons because they are brought to the same temperature as the surrounding coolant.

In contrast to most normal nuclear reactors, however, a fast reactor uses a coolant that is not an efficient moderator, such as liquid sodium, so its neutrons remain high-energy. Although these fast neutrons are not as good at causing fission, they are readily captured by an isotope of uranium (U238), which then becomes plutonium (Pu239). This plutonium isotope can be reprocessed and used as more reactor fuel or in the production of nuclear weapons. Reactors can be designed to maximize plutonium production, and in some cases they actually produce more fuel than they consume. These reactors are called breeder reactors.

Breeder reactors are possible because of the proportion of uranium isotopes that exist in nature. Natural uranium consists primarily of U238, which does not fission readily, and U235, which does. Natural uranium is unsuitable for use in a nuclear reactor, however, because it is only 0.72 percent U235, which is not enough to sustain a chain reaction. Commercial nuclear reactors normally use uranium fuel that has had its U235 content enriched to somewhere between 3 and 8 percent by weight. Although the U235 does most of the fissioning, more than 90 percent of the atoms in the fuel are U238--potential neutron capture targets and future plutonium atoms.

Pu239, which is created when U238 captures a neutron, forms U239 and then undergoes two beta decays, happens to be even better at fissioning than U235. Pu239 is formed in every reactor and also fissions as the reactor operate. In fact, a nuclear reactor can derive a significant amount of energy from such plutonium fission. But because this plutonium fissions, it reduces the amount that is left in the fuel. To maximize plutonium production, therefore, a reactor must create as much plutonium as possible while minimizing the amount that splits.

This is why many breeder reactors are also fast reactors. Fast neutrons are ideal for plutonium production because they are easily absorbed by U238 to create Pu239, and they cause less fission than thermal neutrons. Some fast breeder reactors can generate up to 30 percent more fuel than they use.

Creating extra fuel in nuclear reactors, however, is not without its concerns: One is that the plutonium produced can be removed and used in nuclear weapons. Another is that, to extract the plutonium, the fuel must be reprocessed, creating radioactive waste and potentially high radiation exposures. For these reasons, in the U.S., President Carter halted such spent fuel reprocessing, making the use of breeder reactors problematic.

The U.S. constructed two experimental breeder reactors, neither of which produced power commercially. The Enrico Fermi Nuclear Generating Station in Michigan was the first American fast breeder reactor but operated only from 1963 until 1972 before engineering problems led to a failed license renewal and subsequent decommissioning. Construction of the only other commercial fast breeder reactor in the U.S., the Clinch River plant in Tennessee, was halted in 1983 when Congress cut funding. Elsewhere in the world, only India, Russia, Japan and China currently have operational fast breeder reactor programs; the U.K., France and Germany have effectively shut down theirs.

References

- Fast-Breeder-Reactors: radioactivity.eu.com, Retrieved 26, May 2020
- How-do-fast-breeder-react: scientificamerican.com, Retrieved 07, August 2020

Chapter 5

Nuclear Reaction: Fusion and Fission

A nuclear reaction is a process in which two nuclei or nuclear particles collide to produce different products than the initial particles. Nuclear fusion and nuclear fission are two types of nuclear reactions. Fusion is the process where two nuclei combine together and fission is the splitting of a heavy nucleus into two lighter nuclei. Nuclear reactions and their various aspects like fusion, fission, elastic scattering, inelastic scattering, capture reaction, etc. are covered extensively in this chapter.

Nuclear Reaction

Nuclear reactions generate energy in nuclear reactors, and are responsible for the existence of all elements heavier than hydrogen in the universe. Nuclear reactions denote reactions between nuclei, and between nuclei and other fundamental particles, such as electrons and photons.

Nuclear reactions generate energy in nuclear reactors, in stars, and are responsible for the existence of all elements heavier than hydrogen in the universe. Nuclear reactions denote reactions between nuclei, and between nuclei and other fundamental particles, such as electrons and photons. A short description of the conservation laws and the definition of basic physical quantities is presented, followed by a more detailed account of specific cases: (a) formation and decay of compound nuclei; (b)direct reactions; (c) photon and electron scattering; (d) heavy ion collisions; (e) formation of a quark-gluon plasma; (f) thermonuclear reactions; (g) and reactions with radioactive beams. Whenever necessary, basic equations are introduced to help understand general properties of these reactions.

The collision of two nuclei can give place to a nuclear reaction and, similarly to a chemical reaction, the final products can be different from the initial ones. This process happens when a target is bombarded by particles coming from an accelerator or from a radioactive substance. It was in the latter way that Rutherford observed, in 1919, the first nuclear reaction produced in laboratory, $\alpha + {}^{14}_{7}N \rightarrow {}^{17}_{8}O + p$, using α-particles from a ^{214}Bi sample.

Other reactions were induced using α-particles, the only projectile available initially. With the development of accelerators around 1930, the possibilities multiplied by changing the energy and mass of the projectile. Today it is possible to bombard a target with protons of energy greater than 1 TeV (1 TeV = 10^{12} eV = 1.602×10^{-7} joules), and beams of particles as heavy as uranium are available for study of reactions with heavy ions. Sometimes we can have more than two final products in a reaction, as in:

$$p + {}^{14}N \rightarrow {}^{7}Be + 2\alpha$$

$$p + {}^{23}Na \rightarrow {}^{22}Ne + p + n$$

or just one, as in the capture reaction $p + {}^{27}Al \rightarrow {}^{28}Si*$, where the asterisk indicates an excited state, which usually decays emitting γ-radiation. Under special circumstances, more than two reactants is possible. Thus, for example, the reaction $\alpha + \alpha + \alpha \rightarrow {}^{12}C$ can take place in the overheated plasma of stellar interiors.

The initial and final products can also be identical. This case characterizes a process which can be elastic, as imp $+ {}^{16}O \rightarrow p + {}^{16}O$, where there is only transfer of kinetic energy between projectile and target, or inelastic, as in $n + {}^{16}O \rightarrow n + {}^{16}O*$, where part of the kinetic energy of the system is used in the excitation of ${}^{16}O$. Naturally, nuclear reactions are not limited to nuclei. They could involve any type of particle, and also radiation. Thus, the reactions,

$$\gamma + {}^{63}Cu \rightarrow {}^{62}Ni + p,$$

$$\gamma + {}^{233}U \rightarrow {}^{90}Rb + {}^{141}Cs + 2n,$$

are examples of nuclear processes induced by gamma radiation. In the first case a γ-ray knocks a proton off ${}^{63}Cu$ and in the second it induces nuclear fission of ${}^{233}U$, with the production of two fragments and emission of two neutrons.

Unlike a chemical reaction, the resulting products of a nuclear reaction are not determined univocally: starting from two or more reactants there can exist dozens of final products with an unlimited number of available quantum states. As an example, the collision of a deuteron with ${}^{238}U$ can give place, among others, to the following reactions:

$$d + {}^{238}U \rightarrow {}^{240}Np + \gamma,$$

$$\rightarrow {}^{239}Np + n,$$

$$\rightarrow {}^{239}U + p,$$

$$\rightarrow {}^{237}U + t.$$

In the first of them the deuteron is absorbed by the uranium, forming an excited nucleus of ${}^{240}Np$ that de-excites by emitting a γ-ray. The two following are examples of stripping reactions, in which a nucleon is transferred from the projectile to the target. The last one exemplifies the inverse process: The deuteron captures a neutron from the target and emerges out as ${}^{3}H$ (tritium). This is denoted as a pick-up reaction. Another possibility would be, in the first reaction, those ${}^{240}Np$ fissions instead of emitting a γ-ray, contributing with dozens of possible final products for the reaction.

Each reaction branch, with well-defined quantum states of the participants, is known as channel. In $d + {}^{238}U \rightarrow {}^{240}Np + \gamma \rightarrow {}^{239}Np + n \rightarrow {}^{239}U + p \rightarrow {}^{237}U + t$, for the entrance channel $d + {}^{238}U$, there are four possible exit channels. Notice that a different exit channel would be reached if some of the final products were in an excited state. The probability that a nuclear reaction takes place through a certain exit channel depends on the energy of the incident particle and is measured by the cross section for that channel.

Nuclear reactions proceed through many possible distinct mechanisms. For instance, in direct reactions the projectile and the target have an interaction of short duration, with possible exchange of energy or particles between them. Another mechanism involves the fusion of the projectile with the target, the available energy being distributed among all nucleons, forming a highly excited compound nucleus. The decay of the compound nucleus leads to the final products of the reaction.

In high energies collisions, the nuclei fragment and particles that were not initially present are produced (for instance, pions, kaons, etc). The reactions proceed through an intermediate phase in which the nuclear matter is compressed. At very high energies the quarks and gluons inside nucleons become "free" for a short time forming the quark-gluon plasma. The study of high energy reactions with nuclei is very important for a better understanding of what happens during spectacular stellar explosions (supernovae) and in the interior of compact stars, as for instance, neutron stars. The study of nuclear reactions at high energies allows obtaining information on the equation of state of nuclear matter.

Basic Principles

Conservation Laws

Several conservation laws contribute to restrict the possible processes in a nuclear reaction:

- Baryonic number: There is no experimental evidence of processes in which nucleons are created or destroyed without the creation or destruction of corresponding antinucleons. The application of this principle to low energy reactions is still more restrictive. Below the threshold for the production of mesons (~ 140 MeV), no process related to the nuclear forces is capable to transform a proton into a neutron and vice-versa, and processes governed by the weak force (responsible for the βemission of nuclei) are very slow in relation to the times involved in nuclear reactions (~ 10^{-22} to 10^{-16}s). In this way, we can speak separately of proton and neutron conservation, which should show up with same amounts in both sides of a nuclear reaction.

- Charge: This is a general conservation principle in physics, valid in any circumstance. In purely nuclear reactions it is computed making the sum of the atomic numbers, which should be identical, at both sides of the reaction.

- Energy and linear momentum: These are two of the most applied principles in the study of the kinematics of reactions. Through them, angles and velocities are related to the initial parameters of the problem.

- Total angular momentum: Is always a constant of motion. In the reaction,

  $$^{10}B + {}^{4}He \rightarrow {}^{1}H + {}^{13}C,$$

  ^{10}B has I = 3 in the ground state, whereas the α-particle has zero angular momentum. If it is captured in a swave (l_i = 0), the intermediate compound nucleus is in a state with I_c = 3. Both final products have intrinsic angular momenta equal to 1/2. Hence, their sum is 0 or 1. Therefore the relative angular momentum of the final products will be l_f = 2, 3 or 4.

- Parity: Is always conserved in reactions governed by the nuclear interaction. In the previous example, ^{10}B, ^{4}He and the proton have equal parities, while ^{13}C has odd parity. Therefore, if $l_i = 0$, we necessarily have $l_f = 3$. Thus, the orbital momentum of the final products of $^{10}B + ^4He \rightarrow ^1H + ^{13}C$ is determined by the joint conservation of the total angular momentum and of the parity.

- Isospin: This is an approximate conservation law that is applied to light nuclei, where the effect of the Coulomb force is small. A nuclear reaction involving these nuclei not only conserves the z-component of the isospin (a consequence of charge and baryonic number conservation) but also the total isospin T. Reactions that populate excited states not conserving the value of T are strongly inhibited. An example is the reaction $d + ^{16}O \rightarrow \alpha + ^{14}N$, where the excited state 0+, with 2.31 MeV, of ^{14}N is about a hundred times more populated than the 1^+ ground state. Conservation of energy, angular momentum and parity do not impose any prohibition for that channel, whose low occurrence can only be justified by isospin conservation: the ground states of the four participant nuclei in the reaction have all T = 0 and the state 1^+ of ^{14}N has T = 1.

Kinematics

We consider a typical reaction, in which the projectile a and the target A gives place to two products, b and B. This can also be expressed in the notation that we used so far, a + A → b + B, or even in a more compact notation, A(a,b)B.

Often, a and b are light nuclei and A and B, heavy ones; the nucleus b being emitted at an angle θ and its energy registered in the laboratory system. The recoiling nucleus B has usually a short range and cannot leave the target. It is convenient to introduce the Q-value of the reaction which measures the energy gained (or lost) due to the difference between the initial and final masses:

$$Q = (m_a + m_A - m_b - m_B)c^2$$

Using energy and momentum conservation in the reaction, one gets:

$$Q = E_b\left(1+\frac{m_b}{m_B}\right) - E_a\left(1-\frac{m_a}{m_B}\right) - \frac{2}{m_B}\sqrt{m_a m_b E_a E_b}\cos\theta.$$

From this relation one obtains that, when Q is negative, an energy threshold exists for the incident particle, E_t, as a function of the angle θ, below which nuclei b are not observed in that angle:

$$E_t = \frac{-Q_{mB(mB+mb)}}{m_a m_b \cos^2\theta + (m_B + m_b)(m_B - m_a)}.$$

This example shows the power of conservation laws in the analysis of nuclear reactions. Similar to this example, Very useful relations can be derived in the relativistic regime, and/or using other conserved quantities.

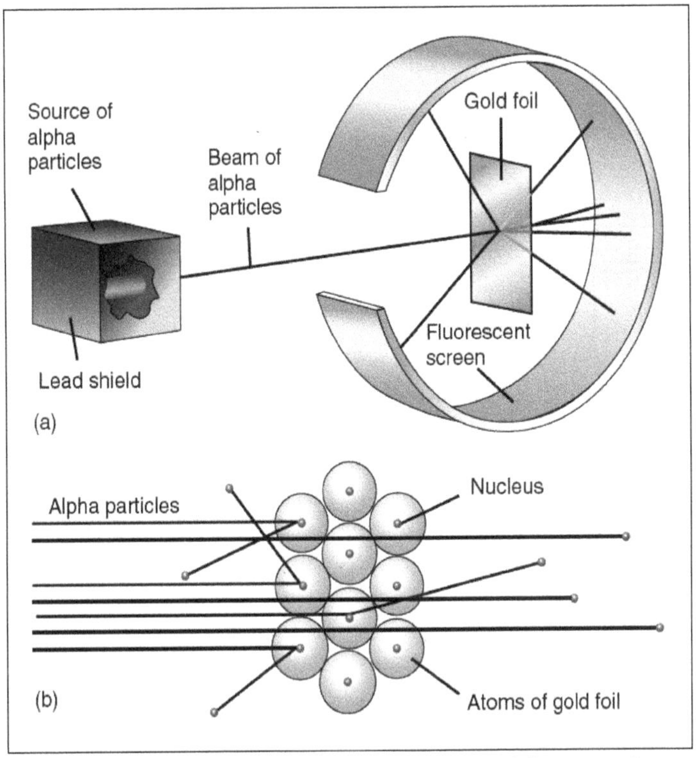

Figure: (Top) Rutherford's scattering experiment with αparticles scattering on a gold foil. (Bottom) Microscopic interpretation of the experimental results.

Cross Sections

Figure shows schematically a typical scattering experiment. In fact this is the sketch of E. Rutherford experiment in 1910. Projectiles (here, α-particles) from a source pass through a collimator and collide with a target (gold foil). Some projectiles are scattered by the target and reach the detector (here, a fluorescent screen). Rutherford expected them to go straight through the target, with perhaps minor deflections. Most did go straight through, but to his surprise some particles bounced straight back! Rutherford hypothesized that the positive alpha particles had hit a heavy mass of positive particles, which he called the nucleus.

Let us consider an experiment as that represented in figure, measuring the count rate of events leading to the population of channel-α, $N_\alpha(\Omega, \Delta\Omega)$. Assuming that interactions between beam particles can be neglected, the count rate $N_\alpha(\Omega, \Delta\Omega)$ should be proportional to the incident flux J, to the fraction of solid angle $\Delta\Omega$ where the particles are scattered to, and to the number of target particles per unit volume. Hence, $N_\alpha(\Omega, \Delta\Omega)$ can then be written as $N_\alpha(\Omega, \Delta\Omega) = (\Delta\Omega \cdot n \cdot J)\, d\sigma_\alpha(\Omega)/d\Omega$. The constant of proportionality,

$$\frac{d\sigma_\alpha(\Omega)}{d\Omega} = \frac{N_\alpha(\Omega, \Delta\Omega)}{\Delta\Omega . n . j}$$

is called the differential cross section for channel-α. This quantity is very useful, since it does not depend on experimental details (detector size, incident flux, target thickness). It depends exclusively on the physics of the projectile and target particles.

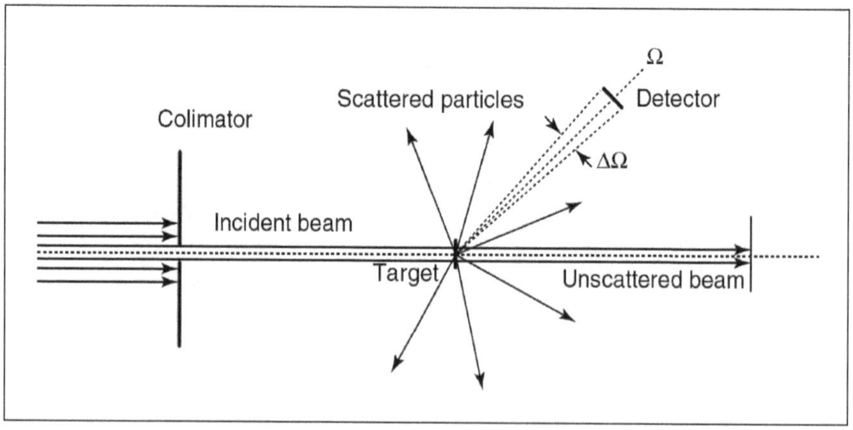

Figure: Schematic representation of a scattering experiment.

Frequently, one is interested in the angle integrated total cross section given by,

$$\sigma_\alpha = \int d\Omega \left[\frac{d\sigma_\alpha(\Omega)}{d\Omega} \right]$$

In nuclear physics cross sections are usually measured in units of barns. 1 barn corresponds to an area of a circle of diameter approximately equal to 8 fm = 8 $\times 10^{-15}$ m. For the comparison between theory and experiment, it is necessary to have cross sections in the same reference frame. Of course, the measured cross section is obtained in the laboratory frame, where the target is at rest. From the theoretical point of view, however, it is important to take advantage of the translational invariance of the projectile-target Hamiltonian, introducing the center of mass frame (CM). The transformation is done using the laws of conservation of energy and momentum.

The Rutherford experiment has a simple classical description in which one assumes that the α-particles follow hyperbolic trajectories in the Coulomb field of the heavy target. The scattering at an angle θ can be easily calculated and with that one is able to calculate the Coulomb, or Rutherford cross section. The result is,

$$\frac{d\sigma_R}{d\Omega} = \left(\frac{Z_p Z_{Te}^{\ 2}}{4E}\right) \cos ec^4 \left(\frac{\theta}{2}\right)$$

where Z_P (Z_T) is the projectile (target) charge number and E is the CM energy.

Elastic Scattering

When a beam of particles, represented in quantum mechanics by a plane wave, hits a nucleus the wave function is modified by the presence of a scattering potential V (r), responsible for the appearance of a phase in the outgoing part of the wave. Elastic scattering is just one of the channels for which the reaction can proceed and is known as elastic channel. Inelastic scattering and all the other channels are grouped in the reaction channel.

The occurrence of a nuclear reaction through a given reaction channel leads to a modification of the outgoing part of the wave function not only by a phase factor, but also by changing its magnitude, indicating that there is a loss of particles in the elastic channel. For a projectile with momentum p =ħ*k*, this can be expressed by,

$$\Psi \sim \frac{1}{2i}\sum_{l=0}^{\infty}(2l+1)i^l P_l(\cos\theta)\frac{\eta_l e^{i(kr-l\pi/2)} - e^{-i(kr-l\pi/2)}}{kr},$$

where the complex coefficient ηl is the factor mentioned above. If η = 1, the sum in equation above can be done analytically, leading to Ψ ~ exp(ik.r), i.e. a plane wave. But if $\eta_l = \exp[i\delta_l]$, with δ_l real, the incoming and outgoing waves have the same magnitude, i.e. the scattering is elastic.

The sum in equation above is known as partial wave expansion of the scattering wave. The label l = 0, 1, 2, · · · = (s, p, d, · · · waves) denotes the contribution of a particular angular momentum (in units of ħ) to the total wavefunction. Classically the angular momentum of an incident particle is given by l = kR, where R is known as impact parameter which is the perpendicular distance to the target if the projectile were undeflected. In quantum mechanics l is not continuous, varying in steps of one.

We can rewrite equation above as a sum of a plane wave and a scattering outgoing wave in the form Ψ ~ exp(ik.r) + f(θ)e^{ikr}/r, where f(θ) accounts for the distortion of the outgoing wave at the scattering angle θ. It is known as the scattering amplitude:

$$f(\theta) = \frac{1}{2ik}\sum_{l=0}^{\infty}(2l+1)(n_l - 1)P_l(\cos\theta)$$

One can now calculate the cross section by counting the number of particles that are scattered through angle θ. This can be done by calculating the particle current associated with the wavefunction Ψ. One gets the differential scattering cross section,

$$\frac{d\sigma_e}{d\Omega} = |f(\theta)|^2 = \frac{1}{4k^2}\left|\sum_{l=o}^{\infty}(2l+1)(1-n_l)P_l(\cos\theta)\right|^2$$

The total scattering cross section, $\sigma_\alpha = \int d\Omega\left[\frac{d\sigma_\alpha(\Omega)}{d\Omega}\right]$ becomes,

$$\sigma_e = \pi\bar{\lambda}^2\sum_{l=0}^{\infty}(2l+1)|1-n_l|^2$$

with $\bar{\lambda} = \lambda/2\pi = 1/k$.

Reaction Cross Sections

To calculate the reaction cross section it is necessary to compute the number of particles that disappear from the elastic channel, what is measured by the flux of the current of probability vector through a spherical surface of large radius centered at the target, calculated with the total wave function of $\Psi \sim \frac{1}{2i}\sum_{l=0}^{\infty}(2l+1)i^l P_l(\cos\theta)\frac{\eta_l e^{i(kr-l\pi/2)} - e^{-i(kr-l\pi/2)}}{kr}$.

$$\sigma_r = \pi\bar{\lambda}^2\sum_{l=0}^{\infty}(2l+1)|1-n_l|^2$$

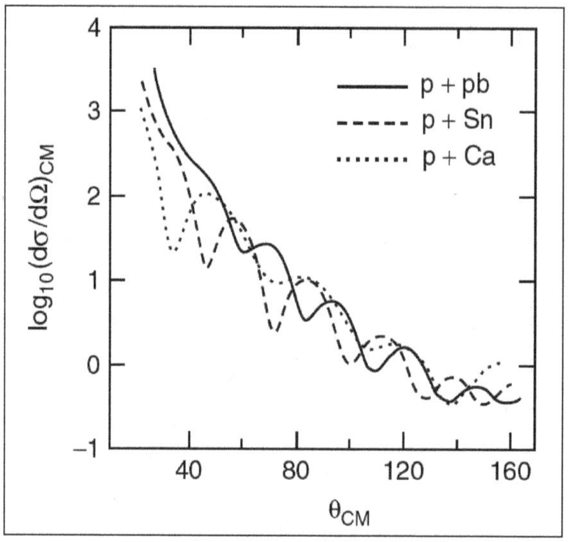

Figure: Elastic scattering angular distribution of protons of 30 MeV on Ca, Sn and Pb.

From $\sigma_e = \pi\bar{\lambda}^2 \sum_{l=0}^{\infty}(2l+1)|1-n_l|^2$ and $\sigma_r = \pi\bar{\lambda}^2 \sum_{l=0}^{\infty}(2l+1)|1-n_l|^2$ when $|\eta_l| = 1$ the reaction cross section is zero and we have pure scattering. The contrary, however, cannot happen, as the vanishing of σ_e also implies in the vanishing of σ_r. In general there is a region of allowed values of η_l for which the two cross sections can coexist.

The maximum of σ_r happens for $\eta_l = 0$, what corresponds to total absorption. Let us suppose that the absorption potential is limited to the surface of a nucleus with radius R $\bar{\lambda}$, that is, that all the particles with impact parameter smaller than the radius R are absorbed. That is equivalent to say that all particles are absorbed for $l \leq R/\bar{\lambda}$. In this case,

$$\sigma_r = \pi\bar{\lambda}^2 \sum_{l=0}^{R/\bar{\lambda}}(2l+1) = \pi(R+\bar{\lambda})^2$$

This is the value that would be intuitively adequate for the total cross section, i.e., equal to the geometric cross section (the part $\bar{\lambda}$ can be understood as an uncertainty in the position of the incident particle). But, we saw above that the presence of scattering is always obligatory.

For $\eta_l = 0$, the scattering and reaction cross sections are identical, yielding the total cross section.

$$\sigma = \sigma_r + \sigma_e = \pi(R+\bar{\lambda})^2 + \pi(R+\bar{\lambda})^2 = 2\pi(R+\bar{\lambda})^2$$

that is twice the geometric cross section.

The presence of the scattering part, that turns the result 18 apparently strange, can be interpreted as the effect of diffraction of the plane waves at the nuclear surface. This effect leads to "shadow" behind the nucleus decreasing its apparent diameter so that, at a certain distance, the perturbation caused by the presence of the nucleus disappears and the plane wave is reconstructed. In this situation we can say that the part of the beam which is diffracted has to be the same as the part that is absorbed, justifying the equality of σ_r and σ_e. The diffraction phenomenon appears clearly in the elastic scattering or inelastic angular distribution (differential cross section as function of the scattering angle): figure exhibits angular distributions for the elastic scattering of 30 MeV protons on ^{40}Ca, ^{120}Sn and ^{208}Pb. The oscillations in the cross sections are characteristic of a Fraunhofer diffraction figure, similar to light scattering by an opaque disk. The angular distance Δθ between the diffraction minima follows closely the expression $\Delta\theta = \hbar/pR$, characteristic of diffraction phenomena.

Excitation Functions

When the projectile has a very low energy, $k \rightarrow 0$, and in particular, $l = kR \ll 1$. As an example consider the scattering of neutrons with l = 0 and ignore the spins of the neutron and of the target. In this case the Schrödinger equation for the radial motion of the neutron with respect to the target is,

$$\frac{d^2u_o}{dr^2}k^2u_0 = 0(r \geq R)$$

This is valid for the radial wave function uo at distances r larger than the channel radius R = Ra + RA, with Ra and RA being the radii of the projectile and of the target, respectively. The solution of equation above is,

$$u_0 = \eta_0 e^{ikr} - e^{-ikr} \qquad (r \geq R)$$

A radial wave function inside the nucleus should connect to the external function 20 with a continuous function and its derivative at r = R. That is, the function,

$$f_l \equiv R\left[\frac{du_l / dr}{u_l}\right]_{r=R}$$

must have identical values if calculated with the internal or the external function and this condition creates a relationship between f_l and η_l. Hence, the knowledge of f_l leads to the knowledge of the cross sections. For neutrons with l = 0, the application of 20 results in,

$$f_0 = ikR\frac{n_o + e^{-2ikR}}{n_o + e^{-2ikR}}$$

from which we extract,

$$n_0 = \frac{f_o + ikR}{f_o - ikR}e^{-2ikR}$$

If f_0 is a real number, then |ηo| 2 = 1. The reaction cross section, $\sigma_r = \pi\bar{\lambda}^2\sum_{l=0}^{\infty}(2l+1)\left|1-n_l\right|^2$, will be zero and we have pure scattering. Using eq, the scattering cross section, equation above, can be written as,

$$\sigma_e, 0 = \pi\bar{\lambda}^2\left|A_{res} + A_{pot}\right|^2$$

with $A_{res} = \frac{-2ikR}{f_0 - ikR}$ and $A_{pot} = \exp(2ikR) - 1$. The separation of the cross section in two parts has physical justification: A_{pot} does not contain f_0 and therefore does not depend on conditions inside the nucleus. It represents the situation where the projectile does not penetrate the nucleus as in the idealized situation where the nucleus is considered an impenetrable hard sphere. The wave function is zero inside the nucleus and u_0 vanishes at r = R, implying $f_0 \to \infty$ and $A_{res} \to 0$. Hence, A_{pot} is the only responsible for the scattering.

Inserting $\sigma_r = \pi\bar{\lambda}^2 \sum_{l=0}^{\infty}(2l+1)\left|1-n_l\right|^2$ in $n_0 = \frac{f_o + ikR}{f_o - ikR}e^{-2ikR}$ and using,

$$f_0 = f_R + if_I$$

We have,

$$\sigma_{r,0} = \pi\bar{\lambda}^2 \frac{-4kR\,f_I}{f_R^2 + (f_I - kR)^2}$$

Statistical Reactions

Compound Nucleus

When a low energy neutron (< 50 MeV) enters the range of nuclear forces it can be scattered or begin a series of collisions with the nucleons. The products of these collisions, including the incident particle, will continue in their course, leading to new collisions and new changes of energy. During this process one or more particles can be emitted and they form with the residual nucleus the products of a reaction that is known as pre-equilibrium. At low energies, the largest probability is the continuation of the process so that the initial energy is distributed among all nucleons, with no emitted particle. The final nucleus with A + 1 nucleons has an excitation energy equal to the kinetic energy of the incident neutron plus the binding energy the neutron has in the new, highly unstable, nucleus. It can, among other processes, emit a neutron with the same or smaller energy to the one absorbed. The de-excitation process is not necessarily immediate and the excited nucleus can live a relatively long time.

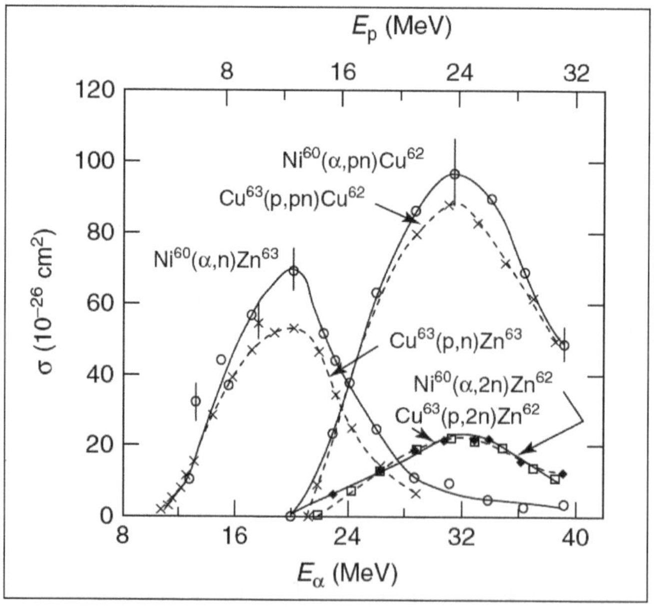

Figure: Cross sections for the reactions shown in 27. The scales of the upper axis (energy of the protons) and lower axis (energy of the α-particle) were adjusted to correspond to the same excitation energy of the compound nucleus.

We say that there is, in this situation, the formation of a compound nucleus as intermediary stage of the reaction. In the final stage the compound nucleus can evaporate one or more particles, fission, etc. In our notation, for the most common situation in which two final products are formed (the evaporated particle plus the residual nucleus or two fission fragments, etc.) we write:

$$a + A \rightarrow C^* \rightarrow B + b,$$

the asterisk indicating that the compound nucleus C is in an excited state.

The compound nucleus lives long enough to "forget" how it was formed and the de-excitation to the final products b and B only depends on the energy, angular momentum and parity of the quantum state of the compound nucleus. An interesting experimental verification was accomplished by S. N. Ghoshal in 1950. He studied two reactions that take to the same compound nucleus, $^{64}Zn^*$, and measured the cross sections of three different forms of decay, as shown below:

$$\begin{array}{ccccc} p + {}^{63}\mathrm{Cu} \searrow & & \nearrow & {}^{63}\mathrm{Zn} + n \\ & {}^{64}\mathrm{Zn}^* & \rightarrow & {}^{62}\mathrm{Cu} + n + p \\ \alpha + {}^{60}\mathrm{Ni} \nearrow & & \searrow & {}^{62}\mathrm{Zn} + 2n \end{array}$$

If the idea of the compound nucleus is valid and if one chooses the energy of the proton and of the incident α-particle to produce the same excitation energy, then the cross section for each one of the three exit channels should be independent of the way the compound nucleus is formed. That is, the properties of the compound nucleus do not have any relationship with the nuclei that formed it. This is confirmed in figure, where one sees clearly that the cross sections depend practically only on the exit channels.

The angular distribution of fragments, or evaporated particles, of a compound nucleus should be isotropic in the center of mass, and this is verified experimentally. However, the total angular momentum is conserved and cannot be "forgotten". Reactions with large transfer of angular momentum, as when heavy ions are used as projectiles, can show a non-isotropic angular distribution in the center of mass system.

The occurrence of a nuclear reaction in two stages allows the cross section for a reaction A(a,b)B to be written as the product, $\sigma(a,b) = \sigma_c(a,A)P(b)$, where $\sigma_c(a, A)$ is the cross section of formation of the compound nucleus starting from the projectile a and the target A and P(b) is the probability that the compound nucleus emits a particle b leaving a residual nucleus B. If the quantum numbers of entrance and exit channels are well specified, i.e., if the reaction begins at an entrance channel α and ends at an exit channel β, one can write,

$$\sigma(\alpha, \beta) = \sigma_c(\alpha)P(\beta)$$

We can associate the probability P(β) to the width Γ_β of the channel β and write:

$$P(\beta) = \frac{\Gamma(\beta)}{T}$$

where Γ is the total width, that is, $\tau = \hbar/\Gamma$ is the halflife of disintegration of the compound nucleus. Equation above just expresses the fact that the decay probability through channel β is the decay rate through that channel divided by the total decay rate. In the competition between the several channels β, the nucleons have clear preference over the γ-radiation whenever there is available energy for their emission and among the nucleons the neutrons have preference as they do not have the Coulomb barrier as an obstacle. Thus, in a reaction where there is no restriction for neutron emission we can say that,

$$\Gamma \cong \Gamma_n,$$

where, Γ_n includes the width for the emission of one or more neutrons.

The study of the function P(β) is done in an evaporation model that leads to results in many aspects similar to the evaporation of molecules of a liquid, with the energy of the emitted neutrons having the form of a Maxwell-Boltzmann distribution,

$$I(E) \propto E\ \exp(-\frac{E}{\theta})dE$$

with I measuring the amount of neutrons emitted with energy between E and E + dE. The quantity θ, with dimension of energy, has the role of a nuclear temperature.

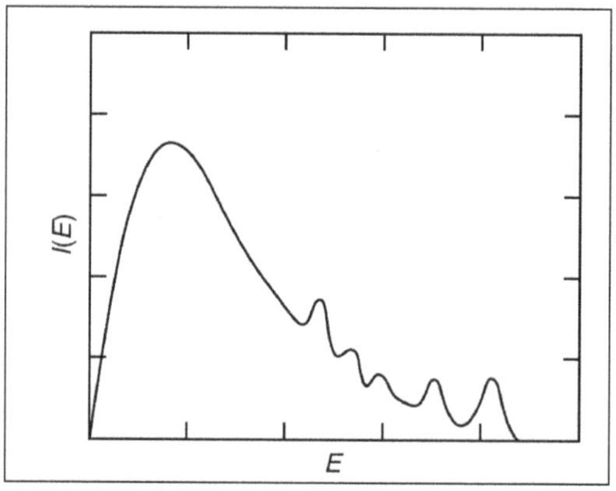

Figure: Energy spectrum of neutrons evaporated by a compound nucleus.

It is related to the density of levels ω of the daughter nucleus B by,

$$\frac{1}{\theta} = \frac{dS}{dE}$$

With,

$$S = \ln \omega(E)$$

where dS/dE is calculated for the daughter nucleus B at the maximum excitation energy that it can have after the emission of a neutron. That is, in the limit of emission of a neutron with zero kinetic energy.

The level density ω (E) is a measure of the number of available energy states for the decay of the compound nucleus in the interval dE around the energy E. In this sense, the relationship 33 is, neglecting the absence of the Boltzmann constant, identical to the thermodynamic relationship between the entropy S and the number of states available for the transformation of a system. Equation above is the well-known relation between the entropy and the temperature.

A simple expression for the energy dependence of the state density is provided by the equidistant spacing model which assumes that the one-particle states are equally spaced with spacing d, and that the total energy of the nucleus is simply obtained by adding the energies of the constituent nucleons. The solution of this problem can be obtained from statistical mechanics.

$$\rho(\mathrm{E}) \sim \exp\left(2\sqrt{aE}\right)$$

Extensive analyzes of experimental data show that for nuclei far from the region of the magic nuclei a varies linearly with A (or with N and Z), as shown in figure.

$$a \cong \frac{A}{k} MeV^{-1}$$

It is found that k ≅ 7.5 – 8.

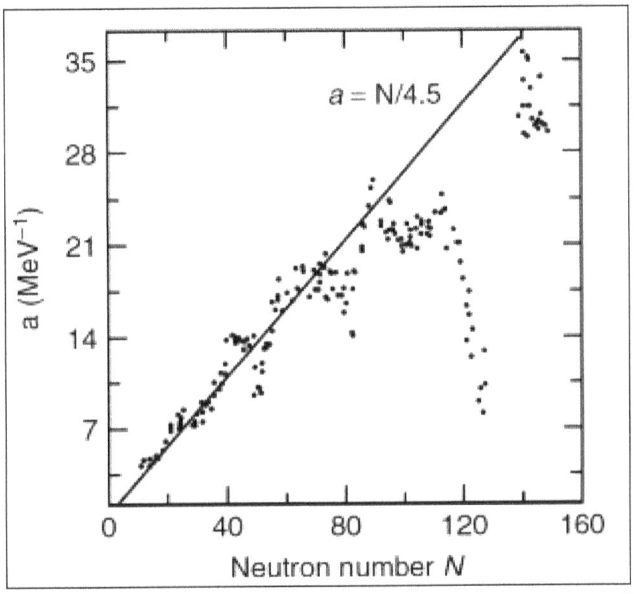

Figure: Values of the level density parameter a as a function of the neutron number. Far from magic regions, a ≈ N/4.5 MeV^{-1} which approximately corresponds to a = A/7.5 MeV^{-1}.

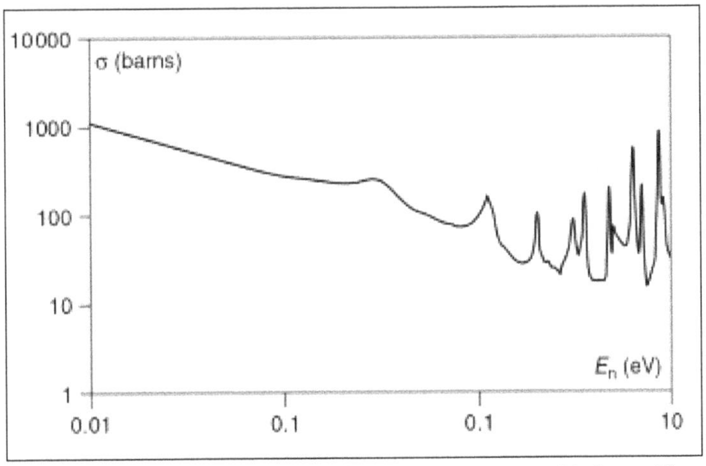

Figure: Total cross section for low energy neutrons hitting ^{235}U.

Energy Spectrum of Neutrons

The energy distribution of neutrons emitted by a compound nucleus has the aspect of the curve shown in figure. Only the low energy part obeys 31 and the reason is simple: the emission of a low energy neutron leaves the residual nucleus with large excitation energy, and the level density is very high. The large density of final states turns the problem tractable with the statistical model that leads to 31. In the opposite situation are the low energy states of the residual nucleus. These isolated states appear as peaks in the tail of the distribution. When the emission is of a proton or of another charged particle, the form of figure is distorted, the part of low energy of the spectrum being suppressed partially by the Coulomb barrier.

Let us now assume that there is only elastic scattering or formation of compound nucleus. We further assume that the elastic scattering is purely potential, without resonant elastic scattering. That is, there is no re-emission of neutrons with energy equal to the incident projectile. This is equivalent to say that the probability that the exit channel is the same as the entrance channel is very low. According to these assumptions, the cross section $\sigma_c(\alpha)$ of $\sigma(\alpha, \beta) = \sigma_c(\alpha)P(\beta)$ is the reaction cross section of $\sigma_r = \pi \bar{\lambda}^2 \sum_{l=0}^{\infty} (2l+1)\left|1-n_l\right|^2$ We can still write the wave function inside the nucleus as just an incoming wave.

$$u_0 \cong \exp(-iKr) \qquad (r < R)$$

where $k=\sqrt{2m(E-V_0)}/\hbar$ is the wavenumber inside the nucleus, and it is assumed that the neutron with total energy E is subject to a negative potential V_0. Equation above is clearly a crude simplification when the incident neutron interacts in a complicated way with the other nucleons in the nucleus. It allows, however, to explain the average behavior of the cross sections for low energies. Starting with $f_l \equiv R\left[\frac{du_l/dr}{u_l}\right]_{r=R}$ and $u_0 \cong \exp(-iKr)$ $(r < R)$ one determines the value of f_0:

$$f_0 = -iKR$$

and from $\sigma_{r,0} = \pi \bar{\lambda}^2 \frac{-4kR\, f_I}{f_R^2+(f_I-kR)^2}$ we get,

$$\sigma_c = \pi \bar{\lambda}^2 \frac{4kK}{(k+K)^2}$$

for the cross section of compound nucleus formation for neutrons with l = 0. At low energies, E << $|V_0|$, thus k << K. Under these conditions, $\sigma_c = 4\pi/kK$, since $\bar{\lambda} = k^{-1} >> R$. Thus, σ_c varies with 1/k. That is,

$$\sigma_c \propto \frac{1}{v}$$

where v is the velocity of the incident neutron. This is the well-known 1/v law that governs the behavior of the capture cross section of low energy neutrons. Figure exhibits the excitation function (cross section as function of the energy) for the reaction $n+^{235}U$. The cross section decays with 1/v up to 0.3 eV, where a series of resonances appear.

Resonances

To understand why there are resonances, we shall use again the simple model of a single particle subject to a square-well potential. We know that inside the well the Schrödinger equation only admits solution for a discrete group of values of energy, E_1, E_2,... E_n. A particle is confined to the interior of the well by reflections that it has at the surface of the well. In these reflections the wave that represents the particle should be in phase before and after the reflection and this only happens for a finite group of energies. Outside the well the Schrödinger equation does not impose restrictions and the energy can have any value. But we know, from the study of the passage of a beam of particles through a potential step, that the discontinuity of the potential at the step provokes reflection even when the total energy of the particles is larger than the step, a situation where classically there would not be any difficulty for the passage of the particles. This reflection is partial and it becomes larger the closer the energy is closer to the height of the potential step. We can say that a particle with energy slightly positive is almost as confined as a particle inside the well. From this fact results the existence of almost bound states of positive energy known as quasi-stationary states or resonances. These resonances appear as peaks in the excitation function, a peak at a given energy meaning that the energy coincides with a given resonance of the nucleus.

The existence of resonances can also be inferred from the properties of the wave function. We consider only elastic scattering, with the other channels closed. The external and internal wavefunctions are both sine functions, the first with wavenumber $k=\sqrt{2m(E-V_o)}/\hbar$ and the second with $k=\sqrt{2m(E-V_o)}/\hbar$. If E is small and Vo is about −35 MeV, we have K >> k. The internal and external parts should join at r = R with continuous function and derivatives. As the internal frequency is much larger than the external one, the internal amplitude is quite reduced. Only at the proximity of the situation in which the derivative is zero there is a perfect matching between both and the internal amplitude is identical to the external one. The energy for which this happens is exactly the energy of resonance.

Resonances appear in the excitation function at relatively low energies, where the number of open channels is not very large and it is possible that to return to the entrance channel. To arrive at an expression of the cross section that describes a resonance, we rewrite 35 as,

$$ul \sim \exp(-iKr) + b\exp(iKr), \qquad (r < R)$$

this time containing a second part which takes into consideration the part of the wave that returns. This second part allows the existence of resonant scattering, where the incident particle is re-emitted with the same energy that it entered, after forming the compound nucleus. The complex amplitude b is always smaller than one, because there are no creation of particles in the region < R in

$$\sigma_{r,0} = \pi\bar{\lambda}^2 \frac{-4kR f_I}{f_R^2 + (f_I - kR)^2}.$$

The second parenthesis in the dWenominator of $\sigma_{r,0} = \pi\bar{\lambda}^2 \frac{-4kR f_I}{f_R^2 + (f_I - kR)^2}$. is never zero, since the numerator forces f_I to be always negative. If for a certain energy f_R vanishes, σ_r,o passes by a maximum in that energy. We can tentatively identify these energies as being the energy of the resonances. Let us take the extreme case of a single resonance at the energy E_R, that is, f_R = o for $E = E_R$. We can expand f_R in a Taylor series in the neighborhood of a resonance, $f_R(E) = (E - E_R)$

$(df_R/dE)E=E_R$ +.... Keeping just the first term of the expansion and using $\sigma_e,0=\pi\bar{\lambda}^2\left|A_{res}+A_{pot}\right|^2$ and $\sigma_{r,0}=\pi\bar{\lambda}^2\dfrac{-4kR\,f_I}{f_R^2+(f_I-kR)^2}$, we get,

$$\sigma_e,0 = \pi\bar{\lambda}^2\left|\exp(2ikR)-1+\frac{i\Gamma_\alpha}{(E-E_R)+i\dfrac{\Gamma}{2}}\right|^2$$

$$\sigma_{r,0} = \pi\bar{\lambda}^2\frac{\Gamma_\alpha(\Gamma-\Gamma_\alpha)}{(E-E_R)^2+\left(\dfrac{\Gamma}{2}\right)^2}$$

where we define:

$$\Gamma\alpha=-\frac{2kR}{(df_R/dE)_{E=E_R}}$$

and

$$\Gamma=\frac{2kR-2f_i}{(df_R/dE)_{E=E_R}}$$

The energy Γ that appears in 40, is the total width of the resonance, Γ = Γα + Γβ +..., i.e., the sum of the widths for all the possible processes of decay of the nucleus, starting from the resonant state. Γα is the entrance channel width, and Γ−Γα is the sum of the widths of all the exit channels except α. If we restrict the exit channels to a single channel β, or we denote β as the group of exit channels except α, equation above is rewritten as:

$$\sigma_{\alpha,\beta} = \pi\bar{\lambda}^2\frac{\Gamma_\alpha\Gamma_\beta}{(E-E_R)^2+\left(\dfrac{\Gamma}{2}\right)^2}$$

which is the usual way of presenting the Breit-Wigner formula, which describes the form of the cross section close to a resonance. Let us recall that equation above refers to an incident particle of l = 0, without charge and without spin. If the spins of the incident and target particles are sa and sA, respectively, and the incident beam is described by a single partial wave l ≠ 0, one can show that the cross section of equation above should be multiplied by the statistical factor g = (2I+1)/ $(2s_a+1)(2s_A+1)$ where I is the quantum number of the total angular momentum $I = s_a + s_A + l$ of the compound nucleus. g reduces, naturally, to the unit in the case of zero intrinsic and orbital angular momenta.

If the exit channel is the same as the entrance channel α, the cross section should be obtained from 40 and its dependence in energy is more complicated because in addition to the resonant scattering there is the potential scattering, and the cross section will contain an interference term between both. The presence of interference results in a peculiar aspect of the scattering cross section, which differs from the simple form 43 for the reaction cross section.

This is seen in figure that shows the forms that a resonance can take in the scattering cross section. The region of energy where resonances show up can extend to 10 MeV in light nuclei but it ends well before this in heavy nuclei. Starting from this limit the increase in level density with energy implies that the average distance between levels is smaller than the width of the levels and individual resonances cannot be resolved experimentally. They form a continuum and this region is known as continuum region. The cross-section in this region fluctuates.

Statistical Theory of Nuclear Reactions

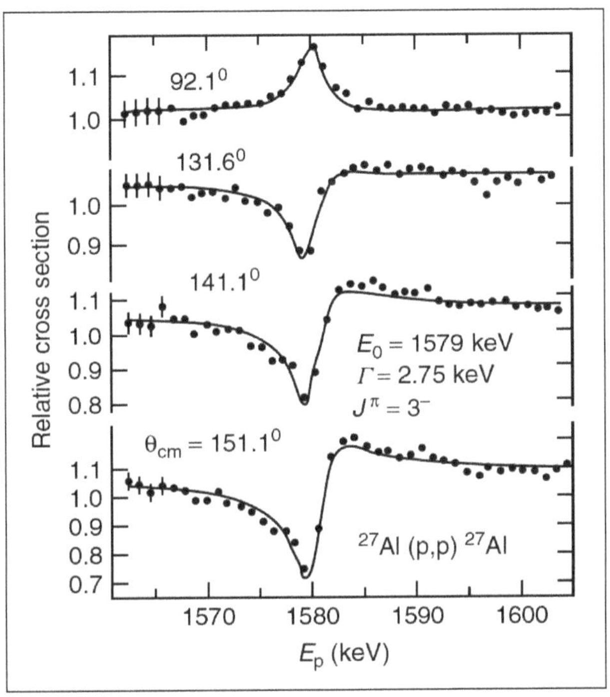

Figure shows differential cross section at four angles for the elastic scattering of protons off ^{27}Al, in units of the Rutherford cross section, in the neighborhood of the 1579 keV resonance. For larger angles we have typical interference between the resonant scattering and the potential scattering.

The fluctuating behavior of low-energy nuclear reactions is due to the interference of the reaction amplitude corresponding to the excitation of each of the overlapping states which vanish in the energy average of the cross-section since these amplitudes are complex functions with random modulus and phase. Calling σN (c) the cross section for the formation of a compound nucleus in the entrance channel c, and using the reciprocity theorem which relates the cross-section σ_{cc}' to the crosssection for the time-reversed process c′ → c, one gets:

$$\sigma_{cc'}(E_{c'})dE_{c'} = \sigma_{CN}(c) \times \frac{(2I_{c'}+1)\mu_{c'}E_{c'}\sigma_{CN}(c')\omega(U_{c'})dU_{c'}}{\sum_c \int_o^{E_c^{max}} (2I_c+1)\mu_c E_c \sigma_{CN}(c)\omega(U_c)dU_c}$$

where I_c is the angular momentum and μ_c is the reduced mass in channel c. Ejectiles with energy in the range $E_{c'}$ to $E_{c'} + dE_{c'}$ leave the residual nucleus with energy in the range $U_{c'}$ to $U_{c'} + dU_{c'}$ where $U_{c'} = E_{CN'} - B_{c'} - E_{c'}$ and E_{CN} and $B_{c'}$ are respectively the compound nucleus energy and the binding energy of the ejectile in the compound nucleus. Equation above 'is the Weisskopf-Ewing formula for the angle-integrated cross-sections. To a good approximation, the level density $\omega(U) \propto \exp(U/T)$, so the ejectile spectrum given by the Weisskopf-Ewing theory is Maxwellian. It rises rapidly above the threshold energy, attains a maximum and then falls exponentially. The Weisskopf-Ewing theory simple to use, but it has the disadvantage that it does not explicitly consider the conservation of angular momentum and does not give the angular distribution of the emitted particles.

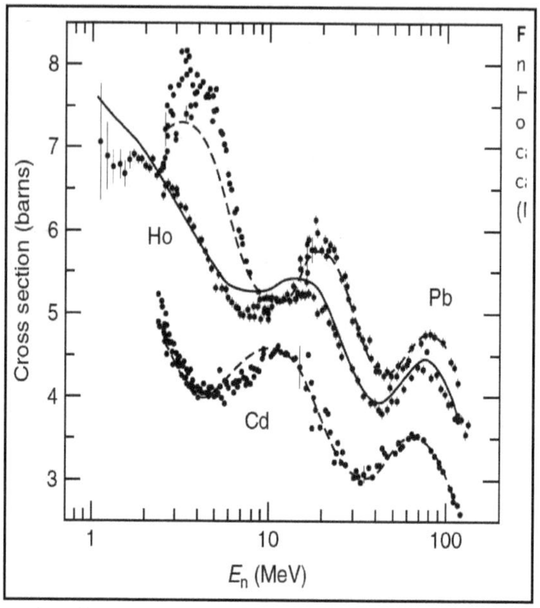

Figure: Total cross section for neutrons incident on cadmium, holmium and lead, showing an oscillatory behavior. The curves for cadmium and lead are results of calculations with the optical model.

This is provided by the Hauser-Feshbach theory. This theory takes into account the formation of the compound nucleus in states of different J and parity π. Let us consider the case of a reaction leading from the initial channel c to a final channel c o. If there is no pre-equilibrium emission, one may identify the compound nucleus formation cross-section $\sigma_{CN} = \sum_{J,\pi} \sigma_{CN}^{J\pi}$ with the optical model reaction cross-section $\sigma_R = \frac{\pi}{k^2} \sum_i (2l+1) T_l$, which, if the transmission coefficients $T_l = 1 - |S_l|^2$ do not depend on J. hSli is the average value of the scattering amplitude over several overlapping resonances.

Using the above assumptions, the reciprocity theorem, and following a similar derivation as with the Ewing-Weisskopf theory, the cross-section for transition c → c′ is given by the Hauser-Feshbach expression,

$$\sigma_{cc'} = \frac{\pi}{k^2} \sum_j \frac{(2J+1)}{(2i_{c+1})(2I_c+1)} \frac{\sum_{8,I} T_{l(c)} \sum_{8',I'} T_{l'(c')}}{\sum_c \sum_{8,l} T_{i(c)}}$$

The compound nucleus states may be both of positive and negative parity. Since parity is conserved, in evaluating, one must take into account that the parity of compound nucleus states and the parity of the residual nucleus states may impose restrictions to the values of the emitted particle angular momentum. Thus, positive parity compound nucleus states decay to positive parity states of the residual only by even angular momenta and to negative parity residual nucleus states by odd angular momenta.

The Optical Model

Expression shows that if a particle is always absorbed when it reaches the nucleus, the total cross section falls monotonically with the energy and grows linearly with $A^{1/3}$. In figure we see curves of total cross section of neutron scattering showing an oscillatory behavior for its energy dependence as well as for the mass dependence of the target. Their presence is mainly due to interference phenomena between the part of the incident beam that passes through the nucleus and the part that passes around it.

The basis of the optical model was established by Herman Feshbach and collaborators in 1953. In this model the interaction between the nuclei in a reaction is described by a potential U(r), with r being the distance between the center of mass of the two nuclei. One replaces the complicated interaction that a nucleon has with the rest of the nucleus with a potential that acts on the nucleon. The potential U(r) includes a complex part that takes into account the absorption effects, i.e., the inelastic scattering. The nuclear scattering is treated in similar form as the scattering of light by a glass sphere and the name of the model derives of this analogy. In its most commonly used form, the optical potential is written as the sum:

$$U(r) = UR(r) + UI(r) + UD(r) + US(r) + UC(r),$$

which contains parameters that can vary with energy and masses of the nuclei and that should be chosen by an adjustment to the experimental data. Obviously, the optical potential U(r) will only make sense if these variations are small for neighboring masses or energies.

The first part of equation above, $U_R(r) = -V f(r, R, a)$, is real and represents a nuclear well with depth $-V$, being multiplied by a Woods-Saxon form factor $f(r, R, a) = [1 + \exp\{(r - R)/a\}]^{-1}$, where R is the radius of the nucleus and a measures the diffuseness of the potential, i.e., the width of the region where the function f is sensibly different from 0 or 1. V, R and a are treated as adjustable parameters.

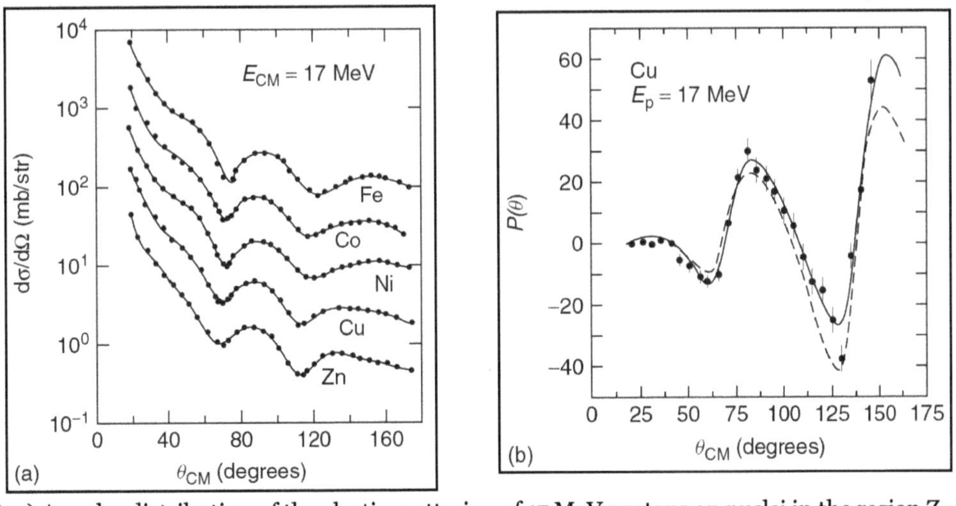

Figure: (Top) Angular distribution of the elastic scattering of 17 MeV protons on nuclei in the region Z = 26 – 30. (Bottom) The polarization value 48 in the scattering of 9.4 MeV protons on copper. The curves are, in both cases, obtained with fits from the optical model.

The absorption effect or, in another way, the disappearance of particles from the elastic channel, is taken into account including the two following imaginary parts U_I (r) = −iW f(r, R_I, a_I) and U_D(r) = 4i$a_I W_D$df(r, R_I, a_I)/dr. An imaginary part produces absorption. It is easy to see this for the scattering problem of the square well: if an imaginary part is added to the well, U(r) = −V_o − iW_o (r < r_o) and U(r) = $_o$ (r > r_o), it appears in the value of K = [2m (E + V_o + iW_o)]1/2/h. This will produce an exponentially decreasing internal wavefunction. Thus, it corresponds to absorption of particles from the incident beam.

U_I is responsible for the absorption in the whole volume of the nucleus, but U_D, built from the derivative of the function f, acts specifically in the region close to the nuclear surface. These two parts have complementary goals: at low energies there are no available unoccupied states for nucleons inside the nucleus and the interactions are essentially at the surface. In this situation U_D(r) is important and U_I (r) can be ignored. On the other hand, at high energies the incident particle has larger penetration and in this case the function U_I (r) is important. As with the shell model potential, used in nuclear structure, a spin-orbit interaction term is added to the optical potential. This term, which is the fourth part of 46, is usually written in the form,

$$\mathrm{U_S}(\mathrm{r}) = \mathrm{s{\cdot}l\,V_s}\frac{1}{r}\frac{d}{dr}f(r, R_S, a_S)$$

s is the spin operator and l the angular orbital momentum operator. As with U_D(r), the part U_S(r) is only important at the surface of the nucleus since it contains the derivative of the form factor f. The values of V_S, R_S and a_S must be adjusted by experiment.

The presence of the term U_S is necessary to describe the effect of polarization. Through experiences of double scattering it can be verified that proton or neutron beams suffer strong polarization at certain angles. This means that the quantity,

$$\mathrm{P} = \frac{N_c - N_b}{N_c + N_b}$$

where N_c is the number of nucleons in the beam with spin up and N_b with spin down, has a value significantly different from zero at these angles. With the inclusion of U_S, the optical model is able to reproduce in many cases the experimental values for the polarization 48.

Finally, a term corresponding to the Coulomb potential is added to 46 whenever the scattering involves charged particles. It has the form,

$$\mathrm{U_C}(\mathrm{r}) = \frac{Z_1 Z_2 e^2}{2R_c}(3 - \frac{r^2}{R_c^2}) \quad (r \le R_c)$$

$$\frac{Z_{1Z_2e^2}}{r} \qquad (r > R_c)$$

where is assumed that the nucleus is a homogeneously charged sphere of radius equal to the Coulomb barrier radius R_c, which separates the regions of nuclear and Coulomb forces.

Figure exhibits the result of the application of U(r) = UR(r) + UI (r) + UD(r) + US(r) + UC (r), to the elastic scattering of 17 MeV protons on several light nuclei. The angular distribution is very well reproduced by the model, which also reproduces correctly the polarization 46 for copper as a function of the scattering angle.

The optical model has a limited set of adjustable parameters and is not capable to describe abrupt variations in the cross sections, as it happens for isolated resonances. However, it can do a good description of the cross sections in the presence of the oscillations of large width in the continuous region, as it treats these as wave phenomena.

Direct Reactions

Direct reactions become more probable as one increases the energy of the incident particle: the wavelength associated to the particle decreases and localized areas of the nucleus can be "probed" by the projectile. In this context, peripheral reactions, where only a few nucleons of the surface participate become important. Direct reactions happen during a time of the order of 10^{-22}s. Reactions with formation of compound nuclei can be up to six orders of magnitude slower. A reaction type at a given energy is not necessarily exclusive; the same final products can be obtained, part of the events in a direct way, other parts through the formation and decay of a compound nucleus.

There are two characteristic types of direct reactions. In the first the incident particle scatters inelastically and the transferred energy is used to excite a collective mode of the nucleus. Rotational and vibrational bands can be studied in this way. The second type involves a modification in the nuclear composition. Examples are transfer of nucleons, as pick-up and stripping reactions. An important reaction of the latter kind is a knock-out reaction where the incident particle knocks out a particle of the target nucleus and continues in its path, resulting in three reaction products. Reactions with nucleon exchange can also be used to excite collective states. An example is a pick-up reaction where a projectile captures a neutron from a deformed target and the product nucleus is in an excited state belonging to a rotational band.

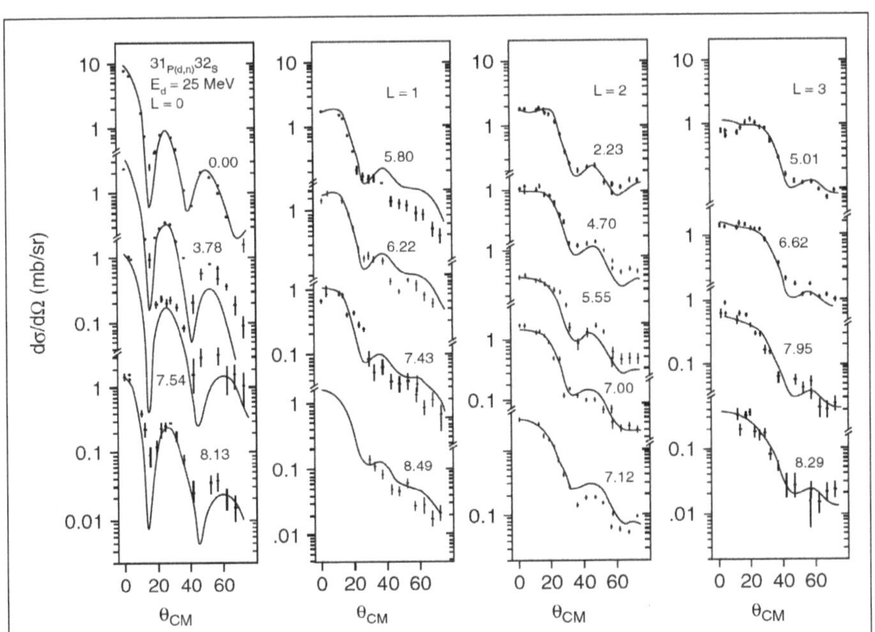

Figure: Angular distribution of the reaction 31P(d,n)32S, with the transfer of a proton to several states of 32S. The curves are results of DWBA calculations for the indicated l values.

Direct reactions exhibit a peculiar form of angular distribution, which allows us to extract information on the reaction mechanism with the employment of simple models. Typical examples are the stripping reactions (d,n) and (d,p), where the angular distribution of the remaining nucleon presents a forward prominent peak and smaller peaks at larger angles, with the characteristic aspect of a diffraction figure.

Figure shows experimental results for the reaction $^{31}P(d,n)^{32}S$: angular distributions of the detected neutrons corresponding to each energy level of ^{32}S are exhibited for several angular momentum. We see that the behavior of the cross sections is in agreement with qualitative predictions: the curves exhibit a first peak at a value of θ that grows with l. Other smaller peaks occur as θ increases. The increase of θ_l with l is an important characteristic that can be used to identify the value of the transferred momentum in a given angular distribution.

We consider an initial quantum state with particles of mass ma hitting a target A of mass m_A. The final quantum states are particles of mass m_b moving away from nucleus B of mass m_B. The differential cross section for the process is given by,

$$\frac{d\sigma}{d\Omega} = \frac{m_a m_b k_b}{(2\pi\hbar^2)^2 k_a} \left|V_{fi}\right|^2$$

involving the matrix element,

$$V_{fi} = \int \Psi_b^* \Psi_B^* \Psi_\beta(r_\beta) V \Psi_a \Psi_A \Psi_\alpha(r_\alpha) d\tau$$

Ψ_a, Ψ_b, Ψ_A, Ψ_B are the internal wavefunctions of the nuclei a, b, A and B. Ψ_α, Ψ_β are the wavefunctions of the relative motion in the entrance channel α and in the exit channel β. The integration volume dτ spans the coordinates of all particles. V is the perturbation potential that causes the "transition" from the entrance to the exit channel.

The use of plane waves for Ψ_α and Ψ_β in equation above is known as first born approximation. With it we can arrive to an approximate expression for the behavior of the differential cross section. As the nuclear forces are of short range, one can restrict the integral 52 to regions where $r_\alpha \cong r_\beta = r$. This leads to,

$$V_{fi} \cong \int dr \exp(ik \cdot r) \left\{ \int \Psi_b^* \Psi_B^* V \Psi_a \Psi_A \, d\tau' \right\}$$

where $k = k_\alpha - k_\beta$. The global variables in dτ were separated into variables dr and dτ′. Expanding the plane wave in a Legendre polynomial series, we obtain,

$$V_{fi} \cong \sum_{l=0}^{\infty} i^l (2l+1) \int j_l(kr) P_l(\cos\theta) F(r) dr$$

where $F(r) = \int \Psi^*_b \Psi^*_B V \Psi_a \Psi_A d\tau'$ contains all the internal properties and is known as as the form factor of the reaction. The action of V is restricted to the surface of the nucleus: outside the nucleus the action of V is limited by the short range of the nuclear forces and inside the nucleus there is a strong deviation to the absorption channel. Expression 54 becomes $V_{fi} \cong \sum_{l=0}^{\infty} c_l j_l(kR)$, where the coefficients cl contain information on the form factor F(r). The index l can be identified as the angular momentum transferred and, for a reaction that involves a single value of l, we can write for the differential cross section:

$$\frac{d\sigma}{d\Omega} \propto |j_l(kR)|^2$$

where the dependence in θ is contained in,

$$k^2 = k_\alpha^2 + k_\beta^2 - 2k_\alpha k_\beta \cos\theta$$

We have an oscillatory behavior for the angular distribution, the maxima separated by π from each other in the axis kR.

The Born approximation with plane waves predicts for certain cases the correct place of the first peaks in the angular distribution but without reproducing correctly the intensities. A considerable progress can be done in the perturbative calculations if, instead of plane waves in 52, we use distorted waves that contain, besides the plane wave, the part dispersed elastically by the optical potential of the target. The Born approximation with distorted waves, or DWBA (distorted wave Born approximation), became a largely employed tool in the analysis of experimental results of direct reactions. With it one can try to extract with a certain reliability the value of the angular momentum l transferred to the nucleus in a stripping or pick-up reaction. An example of this is the already mentioned stripping reaction, ^{31}P (d,n)^{32}S, for deuterons of 25 MeV. For the energy levels shown in figure the assignments of the value of l for the level is, in most cases, univocal.

The angular momentum l transferred in a direct reaction generally modifies the value of the total angular momentum of the nucleus. If J_i is the spin of the target nucleus, the spin J_f of the product nucleus is limited to the values and the initial and final parities obey the relationship $\pi_i \pi_f = (-1)^l$. Equation below allows, with the knowledge of the target nucleus and of the transferred angular momentum, the determination of the parity of the product state formed and is a tool for the determination of its spin.

$$\left| |J_i - l| - \frac{1}{2} \right| \le J_f \le J_i + l + \frac{1}{2}$$

The knowledge of the transferred angular momentum value in a direct reaction opens the possibility to test the predictions of the shell model for the structure of nuclei. In a direct reaction one assumes that the nucleon is located in an orbit of the nucleus with the same angular momentum as the transferred momentum in the reaction. In almost all cases studied with direct reactions the value of the assigned l corresponds exactly to the predicted by the shell model. We know, however, that the real situation is more complicated, due to the presence of the residual interactions that give place to configuration mixing. As result, the cross section for the formation of a state i of the product nucleus is related to that calculated with DWBA for the formation from a single-particle state by,

$$\left(\frac{d\sigma}{d\Omega}\right)_{\exp} = \frac{2J_B + 1}{2J_A + 1} S_{ij} \left(\frac{d\sigma}{d\Omega}\right)_{DWBA}$$

where the spectroscopic factor S_{ij} measures the weight of the configuration j used in the DWBA calculation, in the final state i, with the sum-rule,

$$\sum_j S_{ij} = n_i$$

The sum above embraces all the nucleons in the configurations j of the product nucleus. The statistical weight $(2J_B + 1)/(2J_A + 1)$ that appears in the DWBA calculation involving the angular momentum of the target nucleus JA and final nucleus JB, is explicitly given in equation.

Heavy Ion Reactions

Types and Properties

Heavy ion reactions (with A > 4) can be separated into 3 major categories. Due to their large charge, two heavy nuclei feel a strong mutual Coulomb repulsion. To produce a nuclear reaction the projectile needs enough energy to overcome the Coulomb barrier. For a very heavy target, as $_{238}U$, it is necessary about 5 MeV per nucleon. Then the wavelength of the projectile is small compared with the dimensions of the nuclei and classical and semi-classical methods become useful in the description of the reaction.

The projectile carries a large amount of angular momentum and a good part of it can be transferred to the target in the reaction. Rotational bands with several dozens of units of angular momentum can be created. In fact, heavy ion reactions are the best suited to feed high spin levels.

Direct reactions and formation of compound nucleus are also common processes in reactions with heavy ions. But some peculiarities of these are not found in reactions with projectile nucleons. One of these processes can be understood as intermediate between a direct reaction and the formation of a compound nucleus. Fusion does not occur but projectile and target pass a relatively long time under the mutual action of the nuclear forces. Nuclear matter is exchanged between both and there is a strong heating of the two nuclei, with a large transfer of kinetic energy to the internal degrees of freedom. These are the deep inelastic collisions.

The kind of process that prevails depends upon the distance of closest approach d between the projectile and target. If this distance is sufficiently large only the long range Coulomb interaction acts and, for a classical hyperbolic trajectory, d is related to the impact parameter b and to the energy E of the projectile by $d = \frac{a}{2} + \left[\left(\frac{a}{2}\right)^2 + b^2\right]^{\frac{1}{2}}$ where a is the distance of closest approach in a head-on collision. It is this is related to E by $a = Z_1Z_2e^2/4\pi\varepsilon_0E$.

Experimentally, the variable under control is the energy E of the projectile and, for E sufficiently large, d can be small enough to enter the range of nuclear forces. Collisions near this limit are called grazing collisions and are characterized by values of b_{graz} and d_{graz}. Assuming that there is always reaction when $b < b_{graz}$, the reaction cross section σ_r can be determined geometrically by $\sigma_r = \pi b^2_{graz}$. The experimental determination of σ_r allows to establish the value d_{ras} = 0.5+ 1.36

$(A_1^{1/3} + A_2^{1/3})$ showing that the distance of grazing collision is somewhat larger than that deduced from two touching spheres (1.36 fm > r_0 = 1.2 fm).

When the impact parameter is close to bgraz one expects nuclear reactions of short duration, without the contribution of the compound nucleus formation. Such reactions are elastic and inelastic scattering and transfer of few nucleons. When the incident energy is sufficiently high, small values of b can lead to the projectile penetrating the target. Depending on the energy and on the involved masses, the reaction can end in one of the processes below:

- Fusion: Is the preferred process when one has light nuclei and low energy. There is the formation of a highly excited compound nucleus that decays by evaporation of particles and γ-radiation emission, leading to a cold residual nucleus. If the energy in the CM is close to the Coulomb barrier energy the cross section of compound nucleus formation starting from two nuclei is practically equal to the reaction cross section.

- Fission: When the compound nucleus is heavy the fission process competes strongly with the evaporation of particles in each stage of the evaporation process. A very heavy compound nucleus with large excitation energy has a very small probability of arriving to a cold residual nucleus without fission at some stage of the de-excitation. The role of the angular momentum l transmitted to the target nucleus is also essential. The fission barrier decreases with the increase of l and for a critical value lcrit the barrier ceases to exist. A nucleus with angular momentum greater than lcrit suffers immediate fission and this is also a limiting factor in the production of superheavy elements.

- Deep inelastic collision (DIC): Is a phenomenon characteristic of reactions involving very heavy nuclei (A > 40) and with incident energy of 1 MeV to 3 MeV above the Coulomb barrier. In DIC the projectile and the target spend some time under mutual action, exchanging masses and energy but without arriving to the formation of a compound nucleus. The projectile escapes after transferring part of its energy and angular momentum to its internal degrees of freedom and to the target, with values reaching 100 MeV and 50ħ.

One of the most interesting aspects of DIC is the correlation between the energy dissipated in the collision and the scattering angle in the center of mass. Let us look at figure. The trajectory 1 shows the projectile with an impact parameter that leads it out of the range of nuclear forces. The Coulomb scattering angle will become larger as the impact parameter decreases. In a graph as in figure 12(bottom), where one plots the final energy against the scattering angle, trajectories of type 1 are located in the upper branch, where there is no dissipation and the initial kinetic energy stays unaffected. This upper branch has a maximum value for the scattering angle. At a smaller impact parameter the nuclear attractive force begins to act and, with that the effects of dissipation of DIC. A given Coulomb scattering angle θ can also be reached by the combination of the nuclear and Coulomb forces. Only that now there is loss of energy and the events are located in the branch 2 of figure. There still is no one-to-one correspondence between angle and energy because an infinite number of trajectories can lead to the same angle θ. The branch 2 should be understood as a line of maxima in a three- dimensional representation (called Wilczynski diagram) where the axis perpendicular to the paper is proportional to the cross section $d^2\sigma/dEd\theta$. The same angle θ can also be obtained by the trajectory 3, with a longer interaction time between the nuclei and a larger dissipation. As now the projectile is deflected towards the nucleus, the scattering angle would be formally −θ but, as there is no experimental distinction between θ and −θ, these events appear as an independent ramification in the lower part of the diagram.

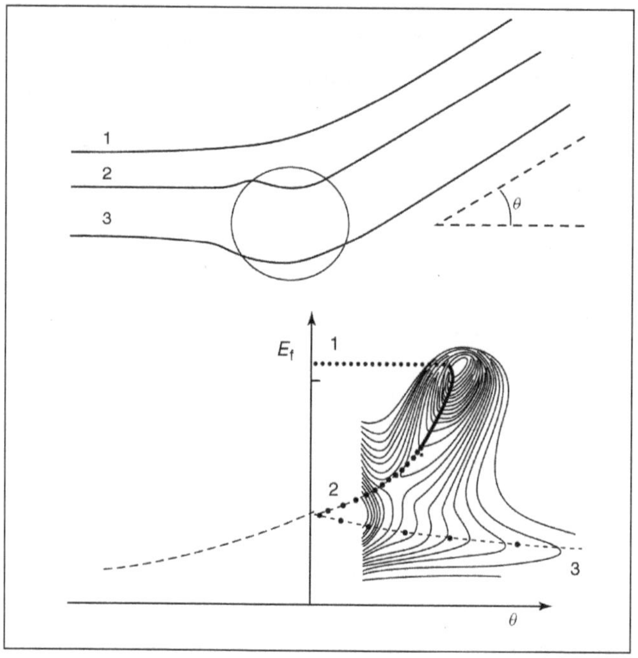

Figure: (Top) Reactions with different impact parameters leading to the same scattering angle θ. (Bottom) Lines of maxima in topographical diagrams of the final energy against the scattering angle.

Superheavy Elements

The heaviest element found in the nature is ${}^{238}_{92}U$. It is radioactive, but it survived since its formation in supernovae explosions because it has a decay half-life of the order of the age of the Earth. Elements with larger atomic number (transuranic) have shorter half-lives and have disappeared. They are created artificially through nuclear reactions using heavy elements as target. Initially, the projectiles used were light particles: protons, deuterons, α-particles and neutrons. The use of neutron is justified because the β^--emission of the compound nucleus increases the value of Z and it was in this way that the first transuranic element, the neptunium, was obtained: n + ${}^{238}U \rightarrow {}^{239}U \rightarrow {}^{239}Np + \beta^-$. Reactions with light particles can produce isotopes up to mendelevium (Z = 101), but it is not possible to go beyond that; the half-lives for α-emission or spontaneous fission become extremely short, turning impracticable the preparation of a target. The alternative is to place a heavy element under the flux of very intense neutrons. This can be done using special reactors or using the rest material of nuclear explosions. The elements einsteinium (Z = 99) and fermium (Z = 100) were discovered in this way in 1955 but the increasing competition that the beta decay has with alpha decay and with spontaneous fission prevents this method to be used for larger Z.

Starting in 1955 heavy ion accelerators began to deliver beams with high enough intensity and energy to compete in the production of transuranic isotopes. The first positive result was the production of two californium isotopes (Z = 98) in the fusion of carbon and uranium nuclei:

$$ {}^{12}_{6}C + {}^{238}_{92}U \rightarrow {}^{244}_{98}Cf + 6n $$
$$ {}^{12}_{6}C + {}^{238}_{92}U \rightarrow {}^{244}_{98}Cf + 4n $$

This opened the possibility of reaching directly the nucleus one wants to create from the fusion of two smaller nuclei. The difficulty of such task is that the cross sections for the production of heavy isotopes are extremely low. As example, the reaction ${}^{50}_{22}Ti + {}^{208}_{82}Pb \rightarrow {}^{257}_{104}Rf + n$ which produces the element rutherfordium, has a cross section of only 5 nb. A small increase in the charges reduces drastically this value: the cross section for the fusion reaction ${}^{58}_{26}Fe + {}^{208}_{82}Pb \rightarrow {}^{265}_{108}Hs + n$ is 4 pb. As comparison, the typical cross sections of DIC for heavy nuclei are in the range 1-2 b.

In spite of the experimental refinement that these low cross sections demand, one is able to produce isotopes with charge as heavy as Z = 118. The understanding of the mechanisms that lead to fusion is, however, not fully understood. According to the traditionally accepted model, the fusion of two nuclei proceeds in two stages: the formation of a compound nucleus and the de-excitation of the compound nucleus by evaporation of particles, preferentially neutrons. The difficulties for the materialization of the process in very heavy nuclei reside in both stages.

Electromagnetic Probes

Coulomb Excitation

Coulomb excitation is an inelastic scattering process in which a nucleus excites another nucleus with its electromagnetic field V. This field can be decomposed in terms of a series of multipoles, e.g. E1, E2, M1, · · ·, which carry well defined angular momenta and parities. At low bombarding energies E2 (quadrupole) excitations are more common, while at higher energies E1 excitations dominate. As an example of a low energy reaction, let us consider the excitation of a quadrupole state in a head-on collision below the Coulomb barrier, i.e. for a situation in which the projectile deccelerates as it approximates the target and stops before reaching the range of the nuclear interaction, reaccelerating backwards after that. The differential cross section is given by the product of the Rutherford differential cross section at 180° and the probability of transition of the target from state i to state f, along the trajectory, measured by the square of,

$$\left.\frac{d\sigma}{d\Omega}\right|_{\theta=180^\circ} = \left.\frac{d\sigma_R}{d\Omega}\right|_{\theta=180^\circ} \times \left|a_{if}\right|^2$$

The square of a_{if} measures the transition probability from i to f and this probability should be integrated along the trajectory. A simple calculation for a_{if} can be done in the case of the excitation from the ground state J = 0 of a deformed nucleus to an excited state with J = 2. The perturbation V comes, in this case, from the interaction of the projectile charge Z_{Pe} with the quadrupole moment of the target nucleus. This quadrupole moment works as an operator that acts between the initial and final states, i.e.

$$V = \frac{1}{2}\frac{Z_{pe^2} Q_{if}}{r^3}$$

where r is the projectile-target separation distance and,

$$Q_{if} = \sum_i \int \Psi^*_f (3z_i^2 - r_i^2) \Psi_i d\tau$$

where the sum extends to all protons at the positions $r_i = (x_i, y_i, z_i)$ in the target. The amplitude is then given by,

$$a_{if} = \frac{4Q_{if}E^2}{3Z_{pe^2\hbar v_o} Z_T^2}$$

Using $\left.\frac{d\sigma}{d\Omega}\right|_{\theta=180^\circ} = \left.\frac{d\sigma_R}{d\Omega}\right|_{\theta=180^\circ} \times \left|a_{if}\right|^2$ for the Rutherford differential cross section at $\theta = 180°$, we obtain,

$$\left.\frac{d\sigma}{d\Omega}\right|_{\theta=180^\circ} = \frac{mE\left|Q_{if}\right|H^2}{18\hbar^2 Z_T^2}$$

which is an expression that is independent of the charge of the projectile. It is, on the other hand, proportional to the mass of the projectile, indicating that heavy ions are more effective for Coulomb excitation.

The quadrupole moment operator Q_{if} uses the wavefunctions Ψ_i and Ψ_f of the initial and final states. If those two wavefunctions are similar, as is the case of an excitation to the first level of a rotational band, the operator Q_{if} can be replaced by the intrinsic quadrupole moment Q. The expression translates, in this way, the possibility to evaluate the quadrupole moment from a measurement of the cross section. This has been indeed a major spectroscopic tool for determining quadrupole momenta (and transitions) along the nuclear chart.

Photonuclear Reactions and Giant Resonances

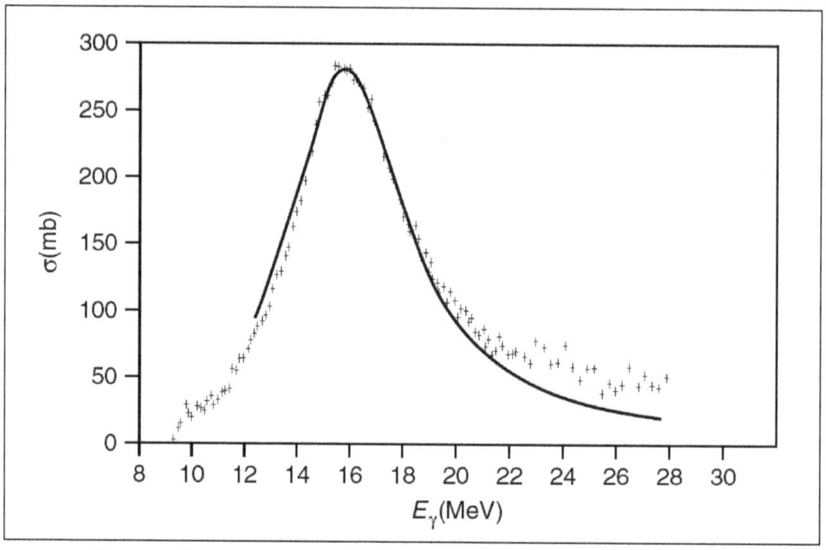

Figure: Giant resonance in the absorption of photons by ^{120}Sn.

A photonuclear reaction is a reaction resulting from the interaction of the electromagnetic radiation with a nucleus. Therefore, one can access information which are complementary to Coulomb excitation (and vice-versa). When the energy of the photon is located above the separation energy of a nucleon, the cross section of photoabsorption reveals the presence of characteristic sharp resonances. But when the incident energy reaches the range of 15-25 MeV, a new behavior appears in the cross section, with the presence of a wide and large peak, called giant electric dipole resonance (E1 excitation). Figure exhibits the excitation function of photo absorption of 120Sn at photon energies around the electric dipole giant resonance at 15 MeV.

The giant resonance occurs in nuclei along the whole periodic table; with the resonance energy decreasing was $E_{GDR} \simeq 80/A^{1/3}$ for A > 20. Their widths are almost all in the range between 3.5 MeV and 5 MeV. In few cases they can reach 7 MeV. They are a collective excitation, i.e., an excitation involving many nucleons at once. The time-varying electric field of the photon is very effective in inducing collective oscillations of protons against neutrons.

The giant electric dipole (GDR) resonance arises from an excitation that transfers by one unit of angular momentum to the nucleus ($\Delta l = 1$). If the nucleus is even even it is taken to a 1− state. The transition also changes the isospin of 1 unit ($\Delta T = 1$) and, due to that, it is also called iso vector resonance. The photon can excite less effectively, but still with appreciable cross sections, giant iso scalar resonances, with $\Delta T = 0$. But electric quadrupole ($\Delta l = 2$) and electric monopole ($\Delta l = 0$) are observed mostly in reactions with charged particles. In a giant electric quadrupole resonance the nucleus oscillates between an spherical (assuming that this is the form of the ground state) to an ellipsoidal form. If protons and neutrons vibrate in phase, we have an isoscalar resonance ($\Delta T = 0$) and if they are oscillate in opposite phase we have an isovector resonance ($\Delta T = 1$). The giant monopole resonance is a very special case of nuclear excitation where the nucleus contracts and expands radially, maintaining its original form but changing its volume. It is also called breathing mode. It can also happen in isoscalar and isovector forms. It is an important way to study the compressibility of nuclear matter. Besides electric giant resonances, associated to a variation in the form of the nucleus, magnetic giant resonances exist, involving spin vibrations. In these, nucleons with spin up move out of phase with nucleons with spin down. The number of nucleons involved in the process cannot be very large because it is limited by the Pauli principle. The magnetic resonances can also separate into isoscalar resonances, where protons and neutrons of same spin vibrate against protons and neutrons of opposite spin, and isovector, where protons with spin up and neutrons with spin down vibrate against their corresponding partners with opposite spins. The last cases, are effectively probed in charge-exchange reactions, i.e. when the projectile charge changes to Z-1 or Z+1, as in (d,p) and (d,n). They are known as giant Gamow-Teller resonances.

Giant resonances can also be produced in an excited nucleus, a case known as the Brink-Axel hypothesis. A special case is when two giant resonances are excited in the same nucleus. The double giant dipole resonance (or multiphonon giant resonance was observed for the first time in reactions with double charge exchange induced by pions in the ^{32}S.

Electron Scattering

Electron scattering experiments have provided a rich database on most stable or long-lived nuclei. Electrons are structureless point-like objects that only interact electromagnetically. Therefore, electron scattering avoids the complexity of the strong interaction between the projectile and the target, and provides clean information about the charge distribution in the nucleus. Considering relativity and the spin of the electrons, the differential cross section for elastic scattering by spinless point-like nuclei can be expressed by the Mott scattering formula (for simplicity, we use ħ = c = 1 units):

$$(\frac{d\sigma}{d\Omega})_{Mott} = \frac{Z_T^2 e^4 \cos^2(\theta/2)}{4p_0^2 \sin^2(\theta/2)\left[1+(2p_o/M)\sin^2(\theta/2)\right]}$$

where Z_{Te} and M are the charge and mass of the target nucleus respectively, and po is the momentum of the incoming electron. Since the nucleus is not a point-like particle, the formula is modified by adding the nuclear electric and magnetic form factors, which contain information of the charge and magnetization distributions inside the nucleus. The Rosenbluth formula accounts for the electron spin and explicitly expresses the cross section for arbitrary nuclei as,

$$(\frac{d\sigma}{d\Omega}) = (\frac{d\sigma}{d\Omega})_{Mott}\{A_o(q^2) +$$

$$B_o(q^2)\left[\frac{1}{2} + (1+\frac{q^2}{4M^2})\right]\tan^2(\theta/2)\ \}$$

where $q^2 = (p_f - p_i)^2 - (E_f - E_i)^2$ is the 4-momentum transfer squared, where p_i (p_f) and E_i (E_f) are the electron incoming (outgoing) momentum and energy. $A_0(q^2)$ and $B_0(q^2)$, are functions of q^2, are the form factors associated with the charge and magnetization distribution of the nucleus respectively. For a spin-0 nucleus:

$$A_o(q^2) = \frac{G_E^2(q^2)}{1+q^2/4M^2} \text{ and } B_0(q^2) = 0$$

where $G_E(q^2)$ is the Sachs charge form factor, and the factor $1 + q^2/4M^2$ is the kinematical recoil correction. For $q \rightarrow 0$, $q^2 \approx q^2$, where q is the three-momentum transfer, and to leading order in powers of q:

$$G_E\left(q^2\right) \simeq G_E(q^2) = \int p\mathrm{T}(\mathrm{r})e^{iq\cdot r}d^3r$$

One thus sees that electron scattering is related to the Fourier transform of the nuclear charge density, ρT (r). If the function GE (q^2) is mapped for a sufficient number of momentum transfers q 2 the Fourier transform can be inverted and ρT (r) can be mapped with precision. For heavy elements (A > 10), this technique shows that the charge radius is closely proportional to 1.12 $A^{1/3}$ fm, where A is the nuclear mass number. This indicates the density saturation for the nuclear matter.

When $q^2 = 0$, inelastic electron scattering probes the same multipolarity transitions (E1, E2, · · ·) as in Coulomb excitation or with real photons. However, in contrast to the later probes, electron scattering also provides information on nuclear excitations for cases in which $q^2 \neq 0$. This is useful in many aspects. For example, at high momentum transfers, the momentum transfer q can be absorbed by a nucleon in the nucleus which acquires an energy of the order of $q^2/2m^*_N$, thus probing the effective mass m^*_N of a nucleon in the nucleus. This is of relevance to understand the effects of nucleon-nucleon interactions in the nuclear environment. The q-region where this occurs is termed by quasi-free scattering. As electrons can penetrate the nuclei, they can also more effectively probe monopole (or breathing mode) excitations of the nuclei, a case in which the nuclear mass distribution vibrates radially. This is of importance for determining the compressibility modulus of the nuclear matter. Finally, in high-energy electron scattering the electron can penetrate deeply inside the nucleons and probe the spin and charge distributions of quarks and gluons inside the nucleon.

Relativistic Nuclear Collisions

Transport Theories and Equation of State

As the bombarding energy in nucleus-nucleus collisions increases the structure aspects of the nuclei become less relevant. Except for the bulk properties of the nuclei (size and number of nucleons), the physics involved is primarily due to the individual, and sometimes collective, hadronic collisions. Several theoretical tools are used to describe these reactions, and in particular we quote (a) time-dependent Hartree-Fock (TDHF), (b) antisymmetrized molecular dynamics (AMD), (c) BoltzmannUehling-Uhlenbeck (BUU, etc.

At intermediate energies of $E_{lab} \sim 100 - 1000$ MeV/nucleon the nucleons and the products of their collisions can be described individually and their propagation can be described by semiclassical equations. One of such equations, and perhaps the most popular in such studies, is the so-called Boltzmann-Uehling-Uhlenbeck (BUU) equation:

$$\frac{\partial f}{\partial t}+\left(\frac{P}{m}+\nabla_p U\right)\cdot\nabla_r f-\nabla_r U\cdot\nabla_r f=$$
$$\int d^3p_2\int d\Omega\sigma_{NN}(\Omega)\left|V_1-V_2\right|$$
$$\times\left\{f_1'f_2'[1-f_1][1-f_2]-f_1f_2\left[1-f_1'\right]\left[1-f_2'\right]\right\}$$

If dN is the number of particles in the volume element d^3 r and whose momenta fall in the momentum element d^3p at time t, then the distribution function f (r, p, t) is given by dN = f (r, p, t) d^3 rd^3p. Thus the BUU equation is an equation for the distribution function f (r, p, t). To account for the effect of each particle interacting with all others, one introduces the concept of mean-field, U (r, p, t). This mean-field exerts a force on each particle, given by $-\nabla_r U$ (r, p, t). Also, the momentum dependence of the potential introduces dependence through the derivative $-\nabla_p U$ (r, p, t).

Due to the nucleon-nucleon collisions, the distribution function within d^3 rd^3p can also be modified by nucleons leaving (or entering) this volume. This is taken care by the collision term, i.e. the right-hand-side of the BUU equation. σNN is the nucleon-nucleon differential scattering cross section, v1 and v2 are the velocities of two colliding nucleons. The first factors inside braces are for collisions populating the volume element and the second term for those depleting it. The factors (1 – f) account for Pauli blocking of final occupied states. The integrals average over scattering angle and over all collisions within d^3 rd^3p. The BUU equation falls in the category of what one calls quantum transport theories. Hadronic transport theories have been quite successful in applications, describing a multitude of measured particle spectra.

Equation above needs as basic ingredients the mean field U and the cross section σ_{NN}. Because these two quantities are related to each other, one should in principle derive them in a self-consistent microscopic approach, as in the fig below.

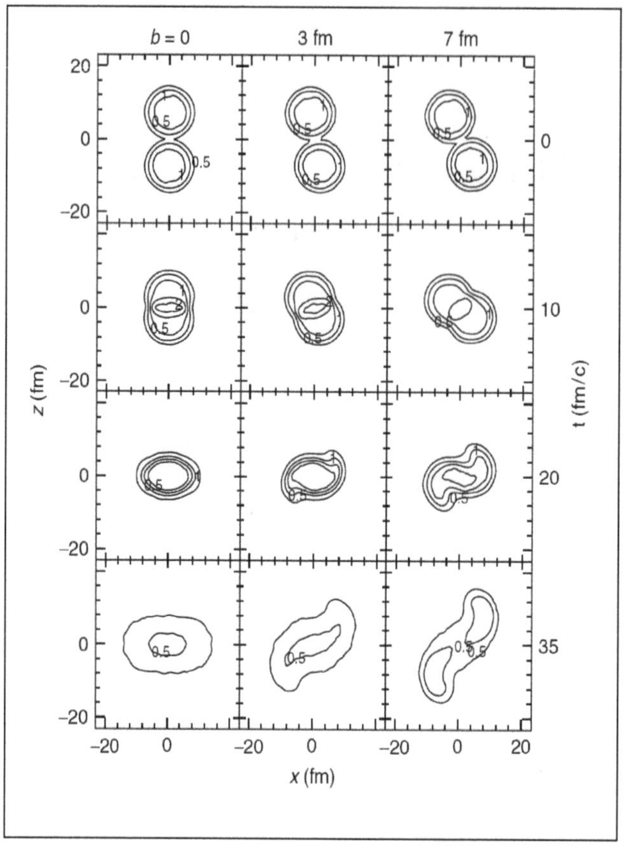

Figuer: Contour plots of baryon density in the reaction plane in Au + Au collisions at 400 MeV/nucleon. The displayed contour lines are for the densities ρ/ρo = 0.1, 0.5, 1, 1.5, and 2.

Brueckner Theory

However, in practice the simulations are often done with a phenomenological mean field and free nuclear cross sections. An important ingredient in the transport theory calculations is the compressibility K of nuclear matter, which refers to the second derivative of the compressional energy E with respect to the density:

$$K = 9p^2 \frac{\partial^2}{\partial p^2}\left(\frac{E}{A}\right)$$

This is an important quantity, e.g., for nuclear astrophysics. Supernova models might or not lead to explosions depending on the value of K. The central collisions of heavy nuclei are one of the few probes of this quantity in the laboratory. The dependence of the calculations on K follow from the dependence of the mean field potential U (U ~ E/A+ kinetic energy terms) on the particle density ρ. A typical parametrization for U is the Skyrme parametrizationU = $a\rho/\rho o + b\,(\rho/\rho o)^{\sigma}$.

The output of eq. is the distribution function f(r, p, t), which allows one to calculate many properties of heavy-ion collisions. Let us quote collective flows, proton and neutron production rates, (sub-threshold and above threshold) pion and kaon yields, etc. Combining eq. with a phase-space coalescence model, one can also calculate such quantities as exclusive flows and intermediate fragment formations.

The dynamics of the central high-energy reactions can be broken down into several stages. Baryon-density contour-plots are shown in fig. for 400 MeV/nucleon Au+Au collisions at b = 0, which will serve to illustrate our points.

Following an initial interpenetration of projectile and target densities, the NN collisions begin to thermalize matter in the overlap region making the momentum distribution there centered at zero momentum in the CM. The density in the overlap region rises above normal and a disk of excited and compressed matter forms at the center of the system. More and more matter dives into the region with compressed matter that begins to expand in transverse directions. At late stages, when the whole matter is excited, transverse expansion predominates.

Kinematics

In relativistic nucleus-nucleus collisions other definitions are best suited for discussing relations between energies, momenta and angles. The rapidity is a variable frequently used to describe the behavior of particles in inclusively measured reactions. It is defined by,

$$y = \frac{1}{2} ln\left(\frac{E + p\|}{E - p\|}\right)$$

which corresponds to tanh $y = \frac{p_{\|}}{E}$, where y is the rapidity, $p_{\|}$ is the longitudinal momentum along the direction of the incident particle, E is the energy, both defined for a given particle. The accessible range of rapidities for a given reaction is determined by the available center-of-mass energy and all participating particles' rest masses. One usually gives the limit for the incident particle, elastically scattered at zero angle:

$$|y_{\max}| = ln\left[\frac{E + p}{m}\right] = ln(\gamma + \gamma\beta)$$

where $\beta \equiv v$ is the velocity and all variables referring to the through-going particle given in the desired frame of reference (e.g. in the center of mass).

Note that $\partial y / \partial p_{\|} = 1/E$. A Lorentz boost β along the direction of the incident particle adds a constant, ln (γ + γβ), to the rapidity. Rapidity differences, therefore, are invariant to a Lorentz boost. Statistical particle distributions are flat in y for many physics production models. Frequently, the simpler variable pseudorapidity η is used instead of rapidity (and sloppy language mixes up the two variables).

$$\eta = -ln\left[\tan\left(\frac{\theta}{2}\right)\right]$$

The pseudorapidity is a handy variable to approximate the rapidity if the mass and momentum of a particle are not known. It is an angular variable defined by whose inverse function is$\theta = 2$ arctan ($e^{-\eta}$), where θ is the angle between the particle being considered and the undeflected beam. η is the same as the rapidity y if one sets β = 1 (or m = 0). Statistical distributions plotted in pseudorapidity rather than rapidity undergoes transformations that have to be estimated by using a kinematic model for the interaction.

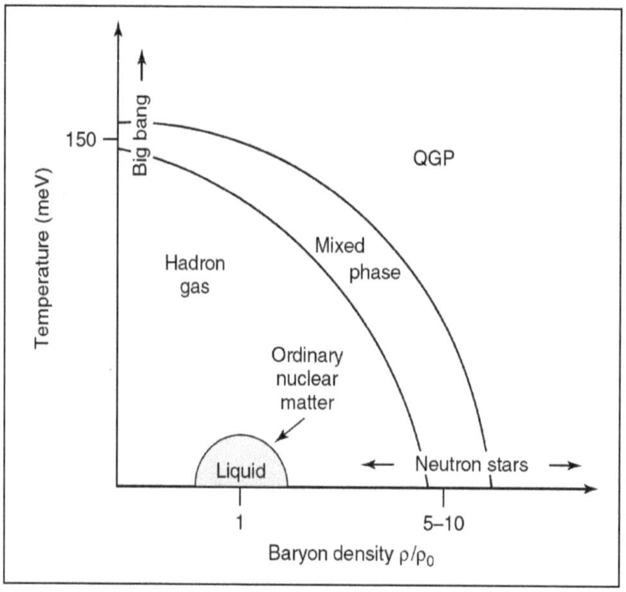

Figure: As water comes in different phases (solid, liquid, gas), so nuclear matter can come in its normal hadronic form and at sufficiently high temperature and density, in the form of a deconfined state of quarks and gluons. The diagram shows how nuclear matter should behave as a function of dentistry and temperature.

The Quark-gluon Plasma (QGP)

The primary motivation for studying ultra-relativistic heavy ion collisions is to gain an understanding of the equation of state of nuclear, hadronic and partonic matter, commonly referred to as nuclear matter. Displayed in fig. is a schematic phase diagram of nuclear matter. The behavior of nuclear matter as a function of temperature and density (or pressure), shown in fig, is governed by its equation of state.

Conventional nuclear physics is concerned primarily with the lower left portion of the diagram at low temperatures and near normal nuclear matter density. Here normal nuclei exist and at low excitation a liquid-gas phase transition is expected to occur. This is the focus of experimental studies using low energy heavy ions. At somewhat higher excitation, nucleons are excited into baryonic resonance states, along with accompanying particle production and hadronic resonance formation. In relativistic heavy ion collisions, such excitation is expected to create hadronic resonance matter.

We now briefly discuss the QGP signatures in nucleus-nucleus collisions. One group of such signatures can be classified as thermodynamic variables. This class involves determination of the energy density ε, pressure P, and entropy density s of the interacting system as a function of the temperature T and the baryochemical potential μB. Experimental observables can be identified with these variables and thus their relative behavior can be determined. If a phase transition to QGP occurs, a rapid rise in the effective number of degrees of freedom, expressed by ε/T^4 or s/T^3, should be observed over a small range of T. The variables T, s, and ε, can be identified with the average transverse momentum $\langle pT \rangle$, the hadron rapidity density dN/dy, and the transverse energy density dE_T/dy, respectively. The transverse energy produced in the interaction is $E_T = \sum_i E_i \sin\theta_i$, where E_i and θ_i are the kinetic energies of the ejectiles and the emission angles.

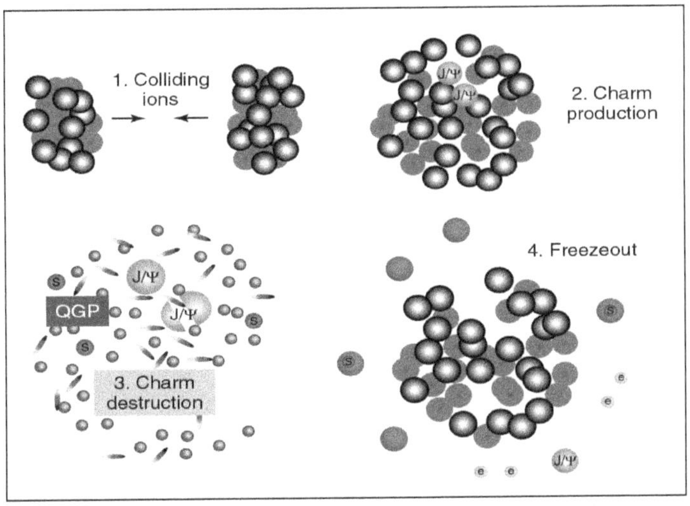

Figure: Formation and evolution of J/Ψ particles in relativistic heavy ion collisions. After the formation, the J/Ψs are dissociated in the plasma due to color screening. The end effect is a smaller number J/Ψs of than expected from pure hadron-hadron multiple collisions.

Electromagnetic (EM) probes, such as photons and leptons, provide information on the various stages of the interaction without modification by final state interactions. These probes may provide a measure of the thermal radiation from a QGP, if a region of photon energy, or equivalently lepton pair invariant mass, can be isolated for emission from a QGP relative to other processes. However, the yields for EM probes are small relative to background processes, which are primarily EM decays of hadrons and resonances. Lepton pairs from the QGP are expected to be identifiable in the 1-10 GeV invariant mass range. The widths and positions of the ρ, ω, and φ peaks in the lepton pair invariant mass spectrum are expected to be sensitive to medium-induced changes of the hadronic mass spectrum.

The production of J/Ψ particles in quark-gluon plasma is predicted to be suppressed. This is a result of the Debye screening of a cc pair, initially formed in the QGP by fusion of two incident gluons. Less tightly bound excited states of the $c\bar{c}$ system, such as Ψ′ and χ_c, are more easily dissociated and will be suppressed even more than the J/Ψ.

A long-standing prediction for a signature of QGP formation is the enhancement of strange hadrons. The production of strange hadrons relative to nonstrange hadrons is suppressed in hadronic reactions. This suppression increases with increasing strangeness content of the hadron. In a QGP the strange quark content is rapidly saturated by ss pair production in gluon-gluon reactions, resulting in an enhancement in the production of strange hadrons. Thus, multi-strange baryons and strange antibaryons are predicted to be strongly enhanced when a QGP is formed.

The connection between energy loss of a quark and the color-dielectric polarizability of the medium can be established in analogy with the theory of electromagnetic energy loss. Although radiation is a very efficient energy loss mechanism for relativistic particles, it is strongly suppressed in a dense medium by the Landau Pomeranchuk effect. Adding the two contributions, the stopping power of quark-gluon plasma is predicted to be higher than that of hadronic matter. A quark or gluon jet propagating through a dense medium will not only loose energy but will also is deflected. This effect destroys the coplanarity of the two jets from a hard parton-parton scattering with the incident beam axis. The angular deflection of the jets also results in an azimuthal asymmetry. The presence of quark-gluon plasma is also predicted to attenuate the emission of jet pairs opposite to the trigger jet.

All of the above quark-gluon plasma signatures have been studied to maturity at the CERN (European Organization for Nuclear Research) - SPS (Super Proton Synchrotron) and the BNL (Brookhaven National Laboratory) - RHIC (Relativistic Heavy Ion Collider) facilities. For more details, the following references discuss: (a) direct photons, (b) dileptons, (c) J/Ψ production and (d) quark in-medium attenuation, also known as high-pT quenching or jet attenuation.

Nuclear Reactions in Stars

Hydrogen and CNO Cycles

Energy production in stars is a well-known process. The initial energy which ignites the process arises from the gravitational contraction of a mass of gas. The contraction increases the pressure, temperature, and density, at the center of the star until values able to start the thermonuclear reactions, initiating the star lifetime. The energy liberated in these reactions yield a pressure in the plasma, which opposes compression due to gravitation. Thus, equilibrium is reached for the energy which is produced, the energy which is liberated by radiation, the temperature, and the pressure.

The Sun is a star in its initial phase of evolution. The temperature in its surface is 6000 °C, while in its interior the temperature reaches 1.5×10^7 K, with a pressure given by 6×10^{11} atm and density 150 g/cm³. The present mass of the Sun is $M_\odot = 2\times10^{33}$ g and its main composition is hydrogen (70%), helium (29%) and less than 1% of more heavy elements, like carbon, oxygen, etc.

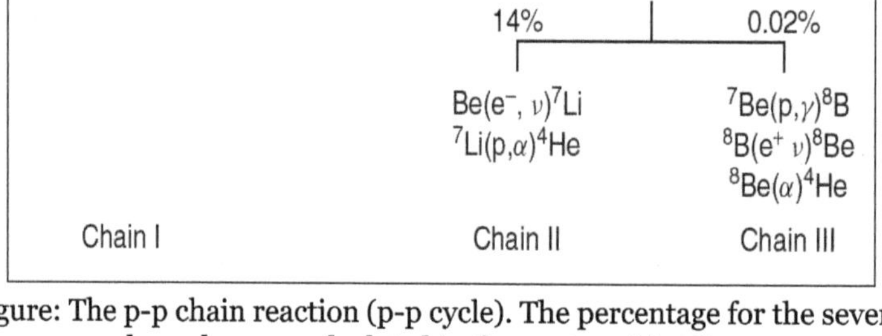

Figure: The p-p chain reaction (p-p cycle). The percentage for the several branches are calculated in the center of the Sun.

What are the nuclear processes which originate the huge thermonuclear energy of the Sun, and that has last 4.6×10^9 years (the assumed age of the Sun)? It cannot be the simple fusion of two protons, or of α-particles, or even the fusion of protons with α-particles, since neither ${}^2_2\text{He}$, ${}^8_4\text{Be}$, or ${}^5_3\text{Li}$, are stable. The only possibility is the proton-proton fusion in the form,

$$p + p \rightarrow d + e^+ + \nu_e,$$

which occurs via β-decay, i.e., due to the weakinteraction. The cross section for this reaction for protons of energy around 1 MeV is very small, of the order of 10^{-23} b. The average lifetime of protons in the Sun due to the transformation to deuterons by means of equation above is about 10^{10} y. This explains why the energy radiated from the Sun is approximately constant in time, and not an explosive process. The deuteron produced in the above reaction is consumed almost immediately in the process,

$$d + p \rightarrow {}_2^3He + \gamma$$

The resulting ${}_2^3He$ reacts by means of:

$${}_2^3He + {}_2^3He \rightarrow {}_2^4He + 2p$$

which produces the stable nucleus ${}_2^4He$ with a great energy gain, or by means of the reaction:

$${}_2^3He + {}_2^4He \rightarrow {}_4^7Be + \gamma$$

In the second case, a chain reaction follows as:

$${}_4^7Be + e^- \rightarrow {}_3^7Li + \nu_{e,} \qquad {}_3^7Li + \mathrm{p} \rightarrow 2({}_2^4\mathrm{He})$$

$${}_4^7Be + p \rightarrow {}_5^8B + \gamma, \qquad {}_5^8B \rightarrow 2({}_2^4He) + e^+ + \nu_e$$

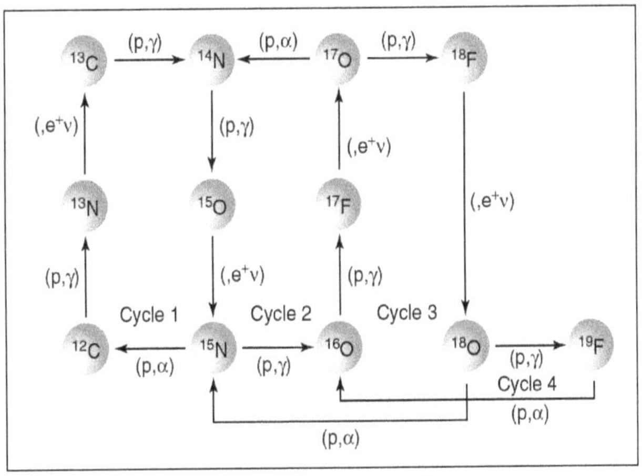

Figure: The CNO cycle.

The chain reaction - is called the hydrogen cycle. The result of this cycle is the transformation of four protons into a α-particle, with an energy gain of 26.7 MeV, about 20% of which are carried away by the neutrinos.

If the star contains heavier elements, another cycle can occur; the carbon cycle, or CNO cycle. In this cycle the carbon, oxygen, and nitrogen nuclei are catalyzers of nuclear processes, with the end product also in the form $4p \rightarrow {}_2^4He$. Fig. describes the CNO cycle. Due to the larger Coulomb repulsion between the carbon nuclei, it occurs at higher temperatures (larger relative energy between the participant nuclei), up to 1.4×10^7 K. In the Sun the hydrogen cycle prevails. But, in stars with larger temperatures the CNO cycle is more effective.

After the protons are transformed into helium at the center of a star like our Sun, the fusion reactions start to consume protons at the surface of the star. At this stage the star starts to become a red giant. The energy generated by fusion increases the temperature and expands the surface of the star. The star luminosity increases. The red giant contracts again after the hydrogen fuel is burned.

Other thermonuclear processes start. The first is the helium burning when the temperature reaches 10^8 K and the density becomes 10^6 g.cm^{-3}. Helium burning starts with the triple capture reaction

$$3({}_{2}^{4}He) \rightarrow {}_{6}^{12}C + 7.65MeV$$

followed by the formation of oxygen via the reaction,

$${}_{6}^{12}C + {}_{2}^{4}He \rightarrow {}_{8}^{16}O + \gamma$$

For a star with the Sun mass, helium burning occurs in about 10^7 y. For a much heavier star the temperature can reach 10^9 K. The compression process followed by burning of heavier elements can lead to the formation of iron and nickel. After that the thermonuclear reactions are no more energetic and the star stops producing nuclear energy.

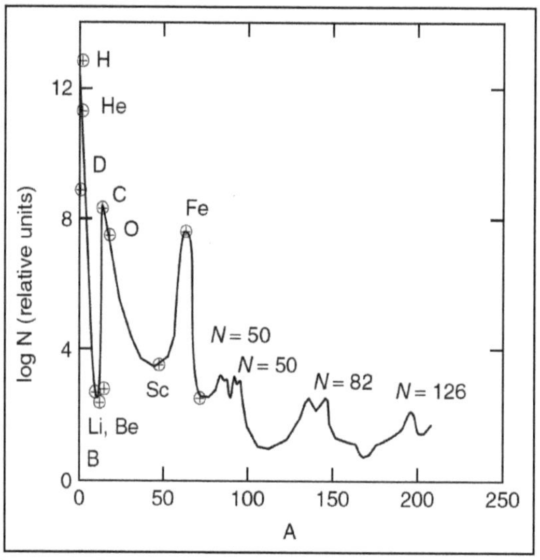

Figure: Relative distribution of elements in our galaxy.

Synthesis of Heavier Elements

In figure we show the relative distribution of elements in our galaxy. It has two distinct regions: in the region A < 100 it decreases with A approximately like an exponential, whereas for A > 100 it is approximately constant, except for the peaks in the region of the magic numbers Z = 50 and N = 50, 82, 126.

The thermonuclear processes - can explain the relative abundance of ${}_{2}^{4}He$, ${}_{6}^{12}C$ and ${}_{8}^{16}O$. The processes occurring after ${}_{2}^{4}He$ burning mainly form isotopes of ${}_{10}^{20}Ne$, ${}_{12}^{24}Mg$ and ${}_{14}^{28}Si$. We can understand the small abundance of the elements Li, Be and B as due to the small velocity with which they are formed via the reaction and the first equation of, while they are rapidly consumed by the second reaction in and the first reaction.

The synthesis of elements heavier than oxygen occur when, after helium burning, a new compression and heating of the star rises the temperature to values higher than 6 × 10⁸ K. This situation ignites an intense carbon burning:

$$^{12}_{6}C + ^{12}_{6}C \rightarrow ^{20}_{10}Ne + ^{4}_{2}He \rightarrow ^{23}_{11}Na + p$$

$$^{23}_{12}Mg + n \rightarrow ^{24}_{12}Mg + \gamma$$

Carbon and oxygen can also burn simultaneously:

$$^{12}_{6}C + ^{16}_{8}O \rightarrow ^{24}_{12}Mg + ^{4}_{2}He,$$

$$^{16}_{8}O + ^{16}_{8}O \rightarrow ^{28}_{14}Si + ^{4}_{2}He.$$

For temperatures above 3×10^9 K more photo-nuclear processes appear. These yield more nuclei to be burned and heavier nuclei are produced:

$$\gamma + ^{28}_{14}Si \rightarrow ^{24}_{12}Mg + ^{4}_{2}He,$$

$$^{4}_{2}He + ^{28}_{14}O \rightarrow ^{32}_{16}S + \lambda.$$

Due to the large number of free neutrons, many (n,γ)- reactions (radiative neutron capture) elements in the mass range A = 28,..., 57 are formed. This leads to a large abundance of elements in the iron mass region, which have the largest binding energy per nucleon. For elements heavier than iron the nuclear fusion processes do not generate energy.

For A > 100 the distribution of nuclei cannot be explained in terms of fusion reactions with charged particles. They are formed by the successive capture of slow neutrons and of β^--decay. The maxima of the element distribution in N = 50, 82, 126 are due to the small capture cross sections corresponding to the magic numbers. This yields a trash of isotopes at the observed element distribution.

Thermonuclear Cross Sections

The nuclear cross section for a reaction between target j and projectile k is defined by,

$$\sigma = \frac{\text{number of reactions target}^{-1}\text{sec}^{-1}}{\text{flux of incoming projectiles}} = \frac{r / n_j}{n_k v}$$

where the target number density is given by n_j, the projectile number density is given by n_k, and v is the relative velocity between target and projectile nuclei. Then r, the number of reactions per cm³ and sec, can be expressed as $r = \sigma v n_j n_k$, or, more generally,

$$r_{j,k} = \int \sigma |v_j - v_k| d^3 n_j d^3 n_k$$

The evaluation of this integral depends on the type of particles and distributions which are involved. For nuclei j and k in an astrophysical plasma, obeying a Maxwell-Boltzmann distribution (MB),

$$d^3 n_j = n_j \left(\frac{m_j}{2\pi kT} \right)^{\frac{3}{2}} \exp(-\frac{m_j v^2_j}{2kT}) d^3 v_j$$

Equation above simplifies to $r_{j,k} = <\sigma v> n_j n_k$, where $<\sigma v>$ is the average of σv over the temperature distribution.

In astrophysical plasmas with high densities and/or low temperatures, effects of electron screening become highly important. This means that the reacting nuclei, due to the background of electrons and nuclei, feel a different Coulomb repulsion than in the case of bare nuclei. Under most conditions (with non-vanishing temperatures) the generalized reaction rate integral can be separated into the traditional expression without screening and a screening factor.

$$<\sigma u>^{*}_{j,k} = f_{scr}(\mathrm{Z_j}, \mathrm{Z_k}, \rho, \mathrm{T}, \mathrm{Y_i}) <\sigma u>_{j,k}$$

This screening factor is dependent on the charge of the involved particles, the density, temperature, and the composition of the plasma. Here Y_i denotes the abundance of nucleus i defined by $Y_i = n_i/(\rho N_A)$, where ni is the number density of nuclei per unit volume and N_A Avogadro's number. At high densities and low temperatures screening factors can enhance reactions by many orders of magnitude and lead to pycnonuclear ignition.

When in $r_{j,k} = \int \sigma |v_j - v_k| d^3n_j d^3n_k$ particle k is a photon, the relative velocity is always c and quantities in the integral are not dependent on d^3n_j. Thus it simplifies to $r_j = \lambda_{j,\gamma} n_j$ and $\lambda_{j,\gamma}$ results from an integration of the photodisintegration cross section over a Planck distribution for photons of temperature T.

$$\mathrm{rj} = \frac{1}{\pi^2 (c\hbar)^3} \int d^3n_j \int_0^{\infty} \frac{c\sigma(E\gamma)E_{\gamma}^2}{\exp(E\gamma/kT)-1} dE_r$$

There is, however, no direct need to evaluate photodisintegration cross sections, because, due to detailed balance, they can be expressed by the capture cross sections for the inverse reaction l + m →j + γ.

A procedure similar to equation above is used for electron captures by nuclei. Because the electron is about 2000 times less massive than a nucleon, the velocity of the nucleus j is negligible in the center of mass system in comparison to the electron velocity ($|v_j - v_e| \approx |v_e|$). The electron capture cross section has to be integrated over a Boltzmann, partially degenerate, or Fermi distribution of electrons, depending on the astrophysical conditions. The electron capture rates are a function of T and $n_e = Y_{ep} N_A$, the electron number density. In a neutral, completely ionized plasma, the electron abundance is equal to the total proton abundance in nuclei $Y_e = P_i Z_i Yi$ and $r_j = \lambda_{j,e}(T, \rho Y_e) n_j$.

This treatment can be generalized for the capture of positrons, which are in a thermal equilibrium with photons, electrons, and nuclei. At high densities ($\rho > 10^{12}$ g.cm^{-3}) the size of the neutrino scattering cross section on nuclei and electrons ensures that enough scattering events occur to thermalize a neutrino distribution. Then also the inverse process to electron capture (neutrino capture) can occur and the neutrino capture rate can be expressed similarly to equation above, integrating over the neutrino distribution. Also inelastic neutrino scattering on nuclei can be expressed in this form. Finally, for normal decays, like beta or alpha decays with half-life τ1/2, we obtain an equation similar to equation above or r_j of the last paragraph with a decay constant $\lambda_j = \ln 2/\tau_{1/2}$ and $r_j = \lambda_j n_j$.

Nuclear Fusion

Nuclear fusion is a reaction through which two or more light nuclei collide into each other to form a heavier nucleus. This reaction takes place with elements that have a low atomic number, such as hydrogen. It is the opposite of nuclear fission reaction in which heavy elements diffuse and form lighter elements. Both nuclear fusion and fission produce a massive amount of energy.

Nuclear Fusion in the Universe

Every star in the universe, including the sun, is alive due to nuclear fusion. It is through this process that they produce such a mind-boggling amount of heat and energy. The pressure at the core of any star is tremendously high and that is where the nuclear fusion reaction takes place.

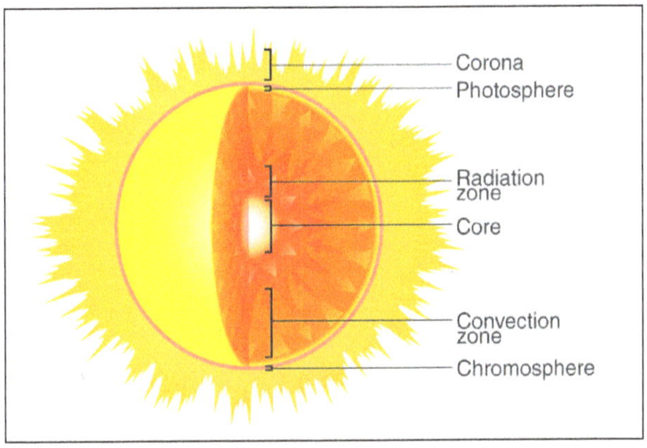

Figure: Example of Nuclear Fusion.

For example, the temperature at the core of the sun is around 15 million degrees Celsius. At this temperature. Coupled with very high pressure, two isotopes of Hydrogen, Deuterium, and Tritium, fuse to form Helium and release the massive amount of energy in the form of heat. Around 600 million tons of hydrogen is converted into Helium every second in the sun.

Difference between Nuclear Fission and Nuclear Fusion

These two are the major nuclear reactions that take place. The basic differences between Nuclear Fission and Nuclear Fusion are:

Nuclear Fission	Nuclear Fusion
Breaks heavy atom into two or smaller ones.	Brings two or more small atoms together to form one large atom.
Does not happen naturally.	The universe is full of instances of nuclear fusion reactions. Every star uses it to produce energy.
Produces a great deal more energy than chemical reactions but still not as much as fusion.	Produces abundant energy than fission reaction.
Does not require a lot of energy to split an atom into two.	Requires a lot of heat and pressure for the process to happen.

Applications of Nuclear Fusion

We are still at an experimental stage as far as nuclear fusion reactions are concerned.

- Clean: No combustion occurs in nuclear power (fission or fusion), so there is no air pollution.
- Less nuclear waste: The fusion reactors will not produce high-level nuclear wastes like their fission counterparts, so disposal will be less of a problem. In addition, the wastes will not be of weapons-grade nuclear materials as is the case in fission reactors.

If utilized properly, nuclear fusion is the answer to the world's power crisis problem. It is clean and produces a minimal amount of nuclear waste as compared to fission reactions. The fuel for fusion, Deuterium, and Tritium, are also readily available in nature. Scientists are hopeful that in the coming centuries, fusion will be a viable alternative power source.

Nuclear Fusion Example

- $H_1^2 + H_1^3 \rightarrow He_2^4 + n_0^1$
- $\Delta E = -17.6$ Me.V.atom^{-1}
- $\Delta E = -1.697 \times 10^9$ kJ.mol^{-1}

Nuclear Fission

Nuclear fission occurs when an atom splits into two or more smaller atoms, most often the as the result of neutron bombardment. Nuclear fission is a process by which the nucleus of an atom is split into two or more smaller nuclei, known as fission products. The fission of heavy elements is an exothermic reaction, and huge amounts of energy are released in the process. The nuclei produced are most often of comparable but slightly different sizes, typically with a mass ratio of products of about 3:2 for common fissile isotopes. Most fission is binary fissions that produce two charged fragments. Occasionally, about 2 to 4 times per 1000 events, three positively charged fragments are produced, which indicates a ternary fission. The smallest of these fragments in ternary processes ranges from the size of a proton to the size of an argon nucleus.

Nuclear fission: In nuclear fission, an unstable atom splits into two or more smaller pieces that are more stable, and releases energy in the process. The fission process also releases extra neutrons, which can then split additional atoms, resulting in a chain reaction that releases a lot of energy. There are also ways to modulate the chain reaction by soaking up the neutrons.

Within the nucleus, there are different forces that act between the particles. The strong nuclear force is the force between two or more nucleons. This force binds protons and neutrons together inside the nucleus, and it is most powerful when the nucleus is small and the nucleons are close together. The electromagnetic force causes the repulsion between like-charged protons. These two forces produce opposite effects in the nucleus. The strong nuclear force acts to hold all the protons and neutrons close together, while the electromagnetic force acts to push protons further apart.

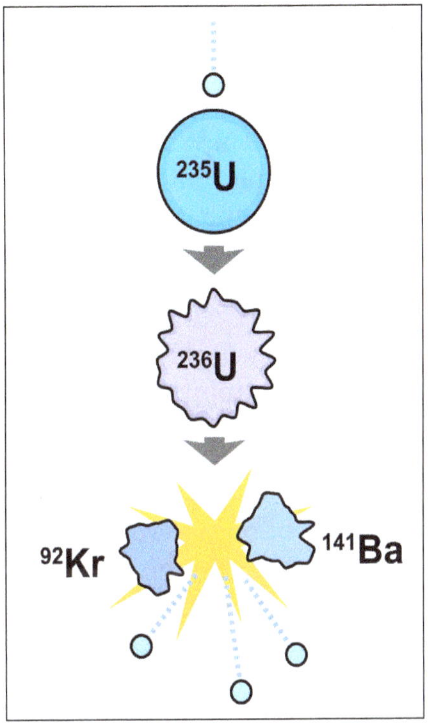

Nuclear fission of U-235: If U-235 is bombarded with a neutron (light blue small circe), the resulting U-236 produced is unstable and undergoes fission. The resulting elements (shown here as Kr-92 and Ba-141) do not contain as many nucleons as U-236, with the remaining three neutrons being released as high-energy particles, able to bombard another U-235 atom and maintain a chain reaction.

In atoms with small nuclei, the strong nuclear force overpowers the electromagnetic force. As the nucleus gets bigger, the electromagnetic force becomes greater than the strong nuclear force. In these nuclei, it's possible for particles and energy to be ejected from the nucleus. These nuclei are called unstable, and this instability can result in radiation and fission.

Neutron Bombardment

In order to initiate fission, a high-energy neutron is directed towards a nucleus, such as ^{235}U. The combination of these two produces ^{236}U, which is an unstable element that undergoes fission. The resulting fission process often releases additional neutrons, which can go on to initiate other ^{235}U atoms, forming a chain reaction. While nuclear fission can occur without this neutron bombardment, in what would be termed spontaneous fission, this is a rare occurrence; most fission reactions, especially those utilized for energy and weaponry, occur via neutron bombardment. If an element can be induced to undergo fission via neutron bombardment, it is said to be fissile.

The Atomic Bomb

Atomic bombs are nuclear weapons that use the energetic output of nuclear fission to produce massive explosions. These bombs are in contrast to hydrogen bombs, which use both fission and fusion to power their greater explosive potential.

Nuclear Chemistry Behind the Explosion

Atomic bombs are made up of a fissile element, such as uranium, that is enriched in the isotope that can sustain a fission nuclear chain reaction. When a free neutron hits the nucleus of a fissile atom like uranium-235 ($_{235}U$), the uranium splits into two smaller atoms called fission fragments, plus more neutrons. Fission can be self-sustaining because it produces more neutrons with the speed required to cause new fissions. This creates the chain reaction.

The uranium-235 content of "weapons-grade" uranium is generally greater than 85 percent, though inefficient weapons, deemed "weapons-usable," can be made of 20 percent enriched uranium. The very first uranium bomb, Little Boy, dropped on Hiroshima in 1945, used 64 kilograms of 80 percent enriched uranium.

In fission weapons, a mass of fissile material, either enriched uranium or plutonium, is assembled into a supercritical mass—the amount of material needed to start an exponentially growing nuclear chain reaction. This is accomplished either by shooting one piece of sub-critical material into another, termed the "gun" method, or by compressing a sub-critical sphere of material using chemical explosives to many times its original density, called the "implosion" method.

The implosion method is considered more sophisticated than the gun method and only can be used if the fissile material is plutonium. The inherent radioactivity of uranium will then release a neutron, which will bombard another atom of $_{235}U$ to produce the unstable uranium-236, which undergoes fission, releases further neutrons, and continues the process. The uranium atom can split any one of dozens of different ways, as long as the atomic weights add up to 236 (uranium plus the extra neutron). The following equation shows one possible split, namely into strontium-95 ($95S_r$), xenon-139 ($139X_e$), and two neutrons (n), plus energy:

$$^{235}U + ^{1}_{0}n \rightarrow ^{95}Sr + ^{139}Xe + 2^{1}_{0} + 180MeV$$

The immediate energy release per atom is about 180 million electron volts (Me). Of the energy produced, 93 percent is the kinetic energy of the charged fission fragments flying away from each other, mutually repelled by the positive charge of their protons. This initial kinetic energy imparts an initial speed of about 12,000 kilometers per second.

However, the charged fragments' high electric charge causes many inelastic collisions with nearby nuclei, and thus these fragments remain trapped inside the bomb's uranium pit. Here, their motion is converted into X-ray heat, a process which takes about a millionth of a second. By this time, the material in the core and tamper of the bomb is several meters in diameter and has been converted to plasma at a temperature of tens of millions of degrees. This X-ray energy produces the blast and fire which are normally the purpose of a nuclear explosion.

Fission bomb assembly methods: Two methods have been applied to induce the nuclear chain reaction that produces the explosion of an atomic bomb. The gun-type assembly uses a conventional explosive to compress from one side, while the implosion assembly compresses from all sides simultaneously.

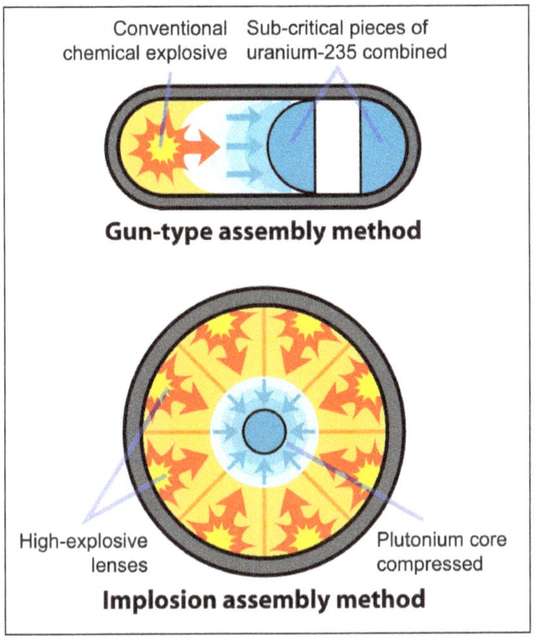

Elastic Scattering

The elastic scattering of a neutron by collision with an atomic nucleus is similar to that of a billiard ball colliding with another billiard ball. A portion of the kinetic energy of one particle is transferred to the other without loss of kinetic energy in the process. In other words, part of the kinetic energy of the neutron can be transferred to a nucleus via collision with the nucleus, and the sum of the kinetic energies of the scattered neutron and recoil nucleus will be equal to the original energy of the colliding neutron. This process of interaction of neutrons with matter results only in scattering of the neutron and recoil nucleus. It does not leave the recoil nucleus in an excited energy state. Elastic scattering is a common mechanism by which fast neutrons lose their energy when they interact with atomic nuclei of low atomic number, such as hydrogen (^{1}H) in light water or paraffin, deuterium (^{2}H) in heavy water, and ^{12}C in graphite, which may be encountered in nuclear reactor moderators. It is easy to conceptualize what would occur when particles of equal or similar mass collide; the event would result in energy transfer and scattering without any other secondary effects, similar to what occurs in billiard ball collisions.

Neutron scattering is the principal mechanism for the slowing of fast neutrons, particularly in media with low atomic number. Let us consider what occurs when a neutron collides with a nucleus and undergoes elastic scattering. Figure illustrates the direction of travel of an incident neutron with given kinetic energy (dashed line). The neutron collides with the nucleus. The nucleus is illustrated as undergoing recoil at an angle β while the neutron is scattered at an angle a to the direction of travel of the incident neutron. The kinetic energy (E_k) lost by the neutron in this collision is defined by the equation.

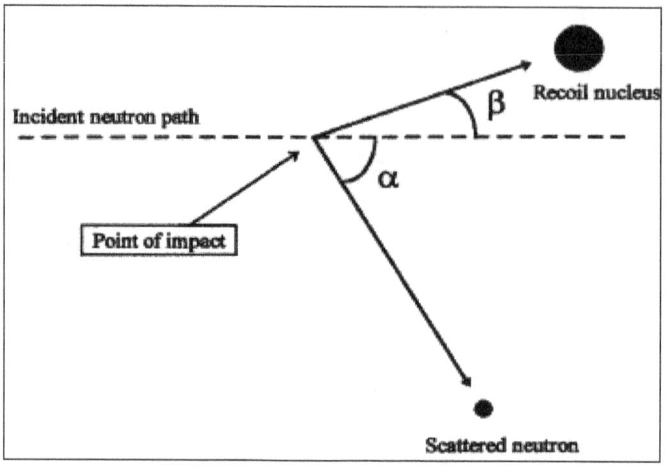

Figure: Elastic scattering of a neutron by collision of the neutron with an atomic nucleus. The neutron is scattered at an angle α and the nucleus recoils at an angle β to the direction of travel of the incident neutron.

$$E_k = \frac{4Mm_m}{\left(M+m_m\right)^2}\cos^2\beta$$

where M is the mass of the nucleus, mn is the mass of the neutron, and β is the recoil angle of the nucleus. A derivation of equation above is provided by Bacon (1969). Let us look at two extreme examples of elastic collisions between a neutron and a nucleus. In the first example, it is intuitively obvious from equation above that for a recoil angle β = 90°, $\cos^2 \beta = 0$ and consequently $E_k = 0$. Under such a circumstance, the neutron is undeflected by the nucleus and there is no energy transfer to the nucleus. The neutron continues along its path undeflected until it encounters another nucleus. For the second case, however, let us consider the other extreme in which the recoil angle, β = 0° where we have a head-on collision of the neutron with the nucleus of an atom. In this case the maximum possible energy of the neutron is imparted to the nucleus, where $\cos^2 \beta = 1$. For example, Table provides the maximum fraction of the kinetic energy that a neutron can lose upon collision with various atomic nuclei. As illustrated in Table, the neutron can transfer more energy to the nuclei of atoms, which have a low mass; and the highest fraction of its energy can be transferred to the nucleus of the proton, which is almost equal in mass to the neutron. Nuclides of low mass number are, therefore, good moderators for the slowing down of fast neutrons. The substances often used are light water (H_2O), heavy water (D_2O), paraffin (C_nH_{2n} + 2), and graphite (C).

Inelastic Scattering

In an inelastic scattering reaction between a neutron and a target nucleus some energy of the incident neutron is absorbed to the recoiling nucleus and the nucleus remains in the excited state. Thus while momentum is conserved in an inelastic collision, kinetic energy of the "system" is not conserved.

In inelastic neutron scattering, the incident neutron is absorbed by the target nucleus and the interaction take place via compound nucleus formation. The compound nucleus emits then a neutron of lower kinetic energy which leaves the original nucleus in an excited state. The nucleus gives up excitation energy by emitting one or more gamma rays to reach its ground state. In comparison to elastic scattering reactions, it is not easy to write an expression for the average energy loss of inelastic scattering reactions because it depends on the energy levels within the target nucleus.

Inelastic scattering occurs above threshold energy. This threshold energy is higher than the energy the first excited state of target nucleus (due to the laws of conservation) and it is given by following formula:

$$Et = ((A+1)/A)^* \varepsilon 1$$

where Et is known as the inelastic threshold energy and ε1 is the energy of the first excited state.

In general, the energy of the first excited state of nuclei decreases with increasing mass number. Therefore an inelastic scattering plays an important role in slowing down neutrons especially at high energies and by heavy nuclei. In calculations of a reactor inelastic scattering by light nuclei (such as oxygen with Et = 6.42 MeV) can ordinarily be ignored. Moreover, the hydrogen nucleus does not have excited states, so only an elastic scattering reaction can occur in that case.

On the other hand an inelastic scattering by uranium nuclei (especially by ^{238}U) must be included. ^{238}U is a major component of nuclear fuel in commercial power reactors, therefore its scattering data are one of the most important data in the neutron transport calculations in the reactor core.

Key Characteristics of Inelastic Scattering

- During an inelastic scattering the neutron is absorbed and then re-emitted.
- The reaction occurs via compound nucleus.
- While momentum is conserved in an inelastic collision, kinetic energy of the "system" is not conserved.
- Some energy of the incident neutron is absorbed to the recoiling nucleus and the nucleus remains in the excited state.
- The nucleus gives up excitation energy by emitting one or more gamma rays.
- General notation: A(n, n')A* or A(n, 2n')B; Example: 14O(n, n')14O*.
- Inelastic scattering is a threshold reaction and occurs above threshold energy.
- Inelastic scattering cross section is relatively small for light nuclei.
- For hydrogen nucleus, inelastic scattering does not occur, because it does not have excited states.
- Inelastic scattering plays an important role in slowing down neutrons especially at high energies and by heavy nuclei (e.g. ^{238}U).

- Inelastic scattering may be significant for heterogeneous reactors with highly enriched fuel (e.g. in fast neutron reactors).

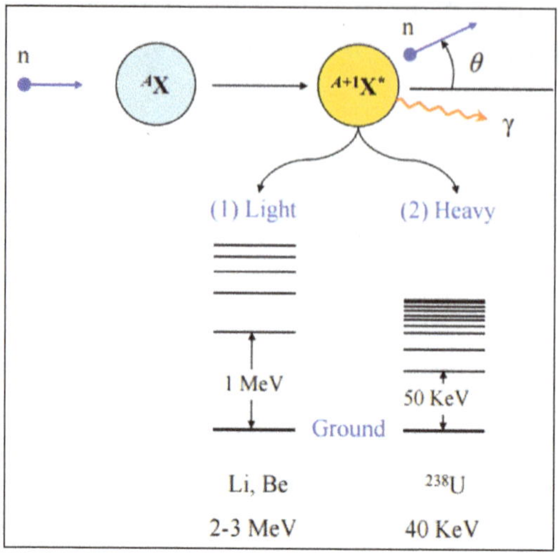

Inelastic Scattering Cross-section

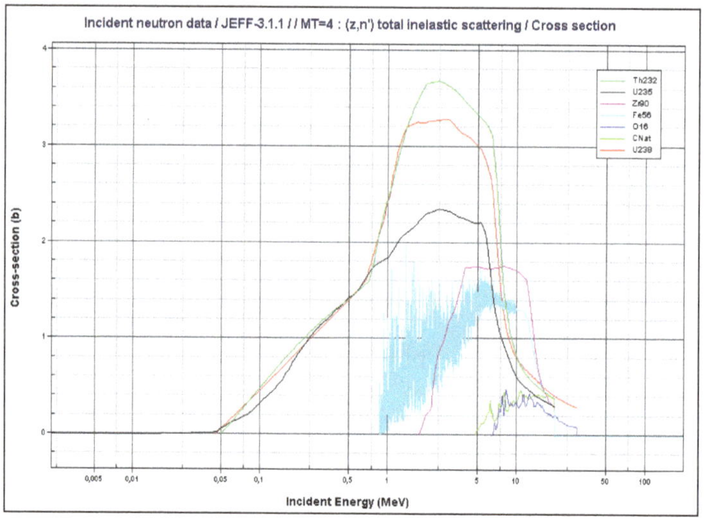

Spallation Reaction

Spallation is a nuclear reaction in which a relativistic light particle like a proton or a neutron hits a heavy nucleus. The energy of the incoming particle usually varies between a few hundred of MeV and a few GeV per nucleon. In a first approximation this interaction process can be grossly divided in two steps.

In the first stage, usually known as intranuclear cascade, the incoming nucleon makes a few, mainly incoherent scattering with nucleons of the target, depositing in this way some fraction of its energy. The incoming nucleon sees the substructure of the nucleus, i. e. a bundle of nucleons, due to the reduced wavelength. This fast stage of nucleon-nucleon scattering interaction leads to the ejection of some of the nucleons and to the excitation of the residual nucleus which will cool itself afterwards.

The de-excitation of the residual nucleus can proceed in two main ways: evaporation and fission. The evaporation is the dedicated de-excitation channel and the excited nucleus emits nucleons or light nuclei such as D, T, He, a, Li, Be. The second important de-excitation mode is fission. In the fission process the nucleus is ultimately "cut" into two fragments of different masses. Generally the fate of the nucleus is its fragmentation.

The nucleus emits particles until its excitation energy goes below the binding energy of the last nucleon. At this state about 8 MeV are remaining and will be emitted by g radiation. The de-excitation process does not end with the ending of the g emission. In fact the nucleus resulting after g decay is often radioactive which will decay until the corresponding stable nucleus is reached. Gaining a better knowledge on the neutron economy may have important consequences on the design of a high-intensity neutron facility. Indeed, if we assume that the neutron economy remains rather stable over a broad range of incident energies, then for a given neutron production the beam intensity could be reduced if higher energies are employed. Let's now have a look to a general description of proton-induced spallation reactions. If the average number of nucleon-nucleon collisions,<n>, is denoted by T, i.e.,

$$T = \langle n \rangle$$

Let's consider an incident proton at impact parameter b(fm), a target nucleus and the z-axis as the beam direction. The average number of nucleon-nucleon collisions taking place between the incident proton and the nucleons of the target is a function of b and can be expressed through the following expression:

$$T(b) = \bar{\sigma}_{NN}\rho(b) = \bar{\sigma}_{NN}\int_{-\infty}^{+\infty}\rho(r)dz$$

where p(r) is the target nuclear matter density and σ_{NN} is the isospin averaged value of the free nucleon-nucleon cross section at the beam energy considered. The probability density of having exactly n nucleon-nucleon collisions can be described by a Poisson distribution around the average value T = <n>:

$$P_n = \frac{T^n e^{-T}}{n!}$$

The cross section σ_n for having n primary collisions at impact parameter b can be calculated as:

$$\sigma_n = 2\pi\int_0^{\infty} p_n b db$$

Adopting this simple model, the mass ejected from the target is proportional to the excitation energy and thus, to the incident proton energy:

$$\langle dm_l \rangle = C(E^*)dT$$

where the slope C is a function of the excitation energy E* and ml is the mass lost from the target ($A-A_F$ where A_F is the mass fragment). The above equation takes into account promptly emitted nucleons (knock-out) since the number of such particles is expected to be proportional to the number of primary nucleon-nucleon collisions. The rather complex kinematics and geometrical situation which governs the emission or retention of struck nucleons in the target and, in addition, the sequential evaporation process which may involve the emission of both nucleons and complex particles, made that the actual mass emitted when an incident proton traverses a target may be subject to considerable fluctuation around the mean. Again is useful to assume a Poisson distribution to describe these fluctuations:

$$P_{ml}(T) = \frac{(CT)^{m!} e^{-CT}}{m_l!}$$

As previously done, this procedure allow to easily write the expression for the cross section σ_{ml}:

$$\sigma_{ml} = \int_0^{\infty} 2\pi b db$$

The analysis carried on until now represent a simple model for proton induced spallation reactions in which a great emphasis has been placed on the necessity of a rather accurate parameterization of nuclear density which permits the calculation of the number of primary nucleon-nucleon collisions. This parameter permits the calculation of the mass evacuated from the target, which follow the energy deposition of the incident proton in the target, and thus an expression for the spallation product cross section.

Simulation Codes

Two widely used transport codes used to simulate the spallation reaction are LCS (LAHET Code System) and FLUKA. The LCS code, developed at Los Alamos National Laboratory is a Monte Carlo code for treating the transport and interactions of nucleons, pions, muons, light ions and antinucleons in complex geometry. LAHET includes both the Bertini and the ISABEL intranuclear cascade model as user options. An evaporation model for the break-up of light nuclei is also included. An optional multistage pre-equilibrium model has been implemented as an intermediate stage between the intranuclear cascade and the evaporation phase of a nuclear reaction. Alternative level density parameterizations are also included. The MCNPX code is another option based on the fully integration of the LAHET code in the MCNP code environment, it only needs of one input file for both codes, avoiding the transfer of large data files.

The FLUKA (FLUctuating KAscade simulation program) code, developed at CERN, is a Monte Carlo code able to simulate transport and interaction of electromagnetic and hadronic particles in any target material over a wide range of energies. FLUKA use the PEANUT (Pre-Equilibrium Approach to Nuclear Thermalization) model to describe an elastic nuclear interaction. This model consists of intranuclear cascade (INC), pre-equilibrium, evaporation and de-excitation. The current version of the code can simulate neutron interaction and transport down to thermal energies (multigroup below 20 MeV) and hadron-hadron and hadron-nucleus interactions up to 100 TeV. The validity of the physical models implemented in FLUKA has been benchmarked against a variety of experimental data over a wide energy range, from accelerator data to cosmic rays data. The FLUKA code has been used for spallation reaction simulation in the Energy Amplifier (EA) project.

Recently a Brazilian research group (IFUSP and CBPF) developed the MCMC/MCEF (Multi-Collisional Monte Carlo plus Monte Carlo for Evaporation-Fission calculation) model to study nuclear reaction such as spallation. The MCMC and the MCEF model utilize the Monte Carlo approach to describe the intranuclear cascade and the evaporation/fission processes respectively. Their coupling originates the CRISP (Colaboração RIo-São Paulo) package. The code takes into account the possibility of neutron, proton and alpha particle evaporation and gives information about neutron and proton multiplicity, angular distribution and energy spectra. Some preliminary results were obtained calculating neutron multiplicities in ^{208}Pb for 200-1200 MeV protons. The CRISP results where compared with results obtained with the LAHET code using both the Bertini and the ISABEL model and with experimental data. A very good accordance with experimental data was registered. The qualification of the CRISP package for ADS calculation is being performed.

Nuclear Decay

Nuclear decay occurs when the nucleus of an atom is unstable and spontaneously emits energy in the form of radiation. The result is that the nucleus changes into the nucleus of one or more other elements. These daughter nuclei have a lower mass and are more stable (lower in energy) than the parent nucleus. Nuclear decay is also called radioactive decay, and it occurs in a series of sequential reactions until a stable nucleus is reached.

Nuclear reactions release much more energy—orders of magnitude more—than exothermic chemical reactions. Nuclear radiation has applications in energy production, weapons development, cancer treatment, and imaging science. The first two applications are often politically fraught.

Figure: Atomic bombs make a distinctive mushroom-shaped cloud.

Neutron Proton Ratio

The neutron-to-proton ratio $n:p$ or $\frac{n}{p}$ determines whether or not an atom is stable. Elements with a ratio close to or equal to one are considered stable. Regardless of the neutron-proton ratio, all elements with an atomic number greater than 83 are unstable and therefore radioactive.

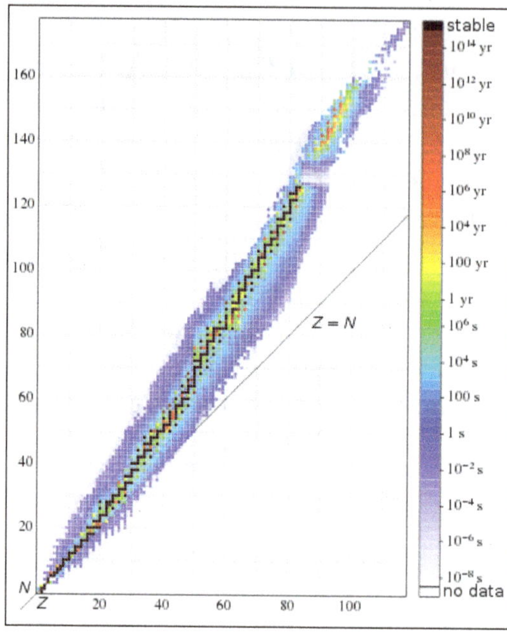

The black line in the picture above shows the stability band, where the ratio is equal to 1:1. As the atomic number increases, the graph curves upward. As atomic mass increases, a larger number of neutrons are required to overcome the strong electrostatic repulsion between the protons. These atoms do not have a stable $\frac{n}{p}$ ratio. To overcome this instability, the nucleus of the atom emits energy. The emissions are called *radioactivity.*

Types of Nuclear Decay

There are six common types of nuclear decay.

Alpha decay produces a helium-4 nucleus, which is also known as an *alpha particle.* The daughter nucleus therefore contains two fewer protons and two fewer neutrons than the parent. This type of emission is commonly observed in nuclei where the atomic mass is 200 or greater.

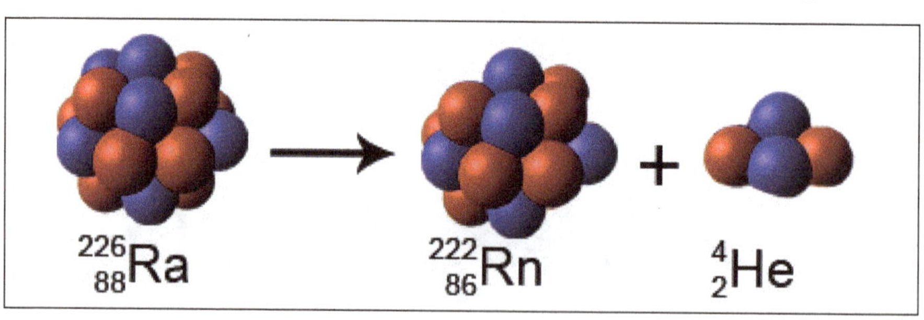

Beta decay is commonly observed in nuclei that have a large number of neutrons. A neutron is split into a proton and a high-energy electron (called the beta particle), the latter of which is ejected from the nucleus.

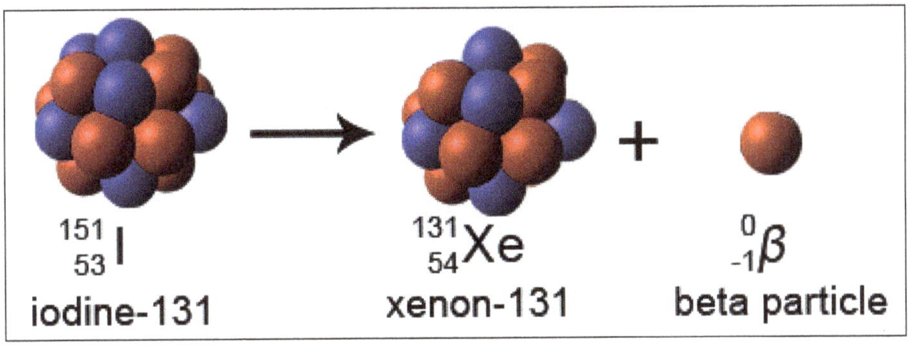

Electron capture occurs when an electron in the inner shell combines with a proton to form a neutron. Once there is an opening in the inner shell, a second electron will move down to a lower energy state, leading to emission of an X-ray.

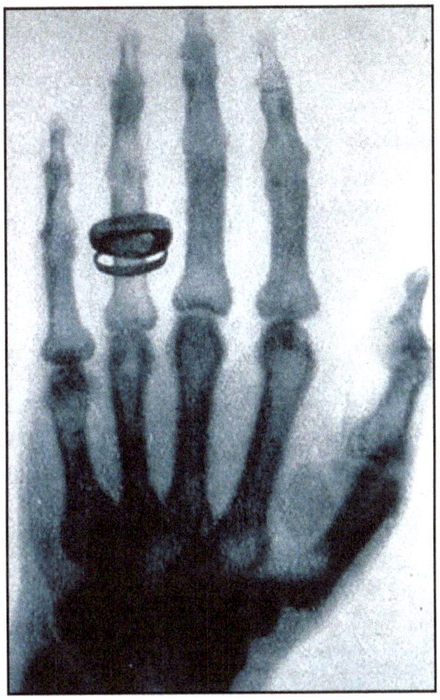

Figure: German physicist Wilhelm Rontgen took this X-ray during a public lecture in 1896.

Gamma emission is unique in that it does not necessarily change one element into another. Often, the products of nuclear decay reactions are formed in an excited state. Similar to the way an electron in an excited state will emit energy as it returns to the ground state, the daughter nuclei release a high-energy photon (a gamma ray) as it reaches its stable form. This process may take place instantaneously or several hours after the first nuclear reaction has taken place, depending on the element.

Positron emission can be thought of as the opposite of beta decay. A proton is split to make a neutron and a positron. (A positron has the same mass as an electron, but the opposite charge.) The positron is then ejected from the nucleus. Positron emission tomography (PET) is commonly used in medicine.

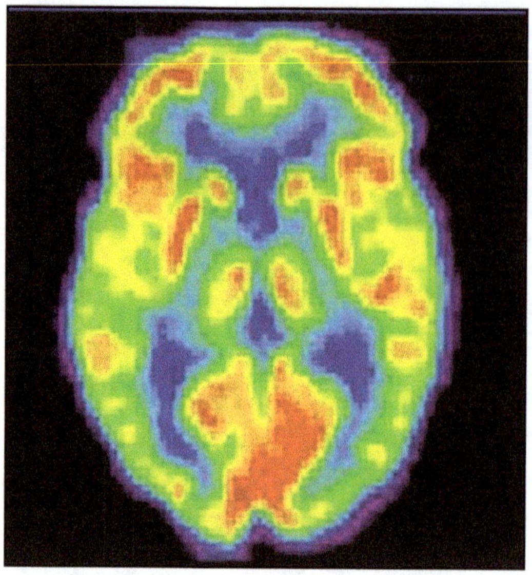

Figure: PET image of a human brain.

Spontaneous fission occurs when a nucleus breaks completely, creating two separate pieces with different atomic numbers and atomic masses. An element must be very massive and have a high neutron-to-proton ratio in order to undergo spontaneous fission. Fission emits a large amount of energy.

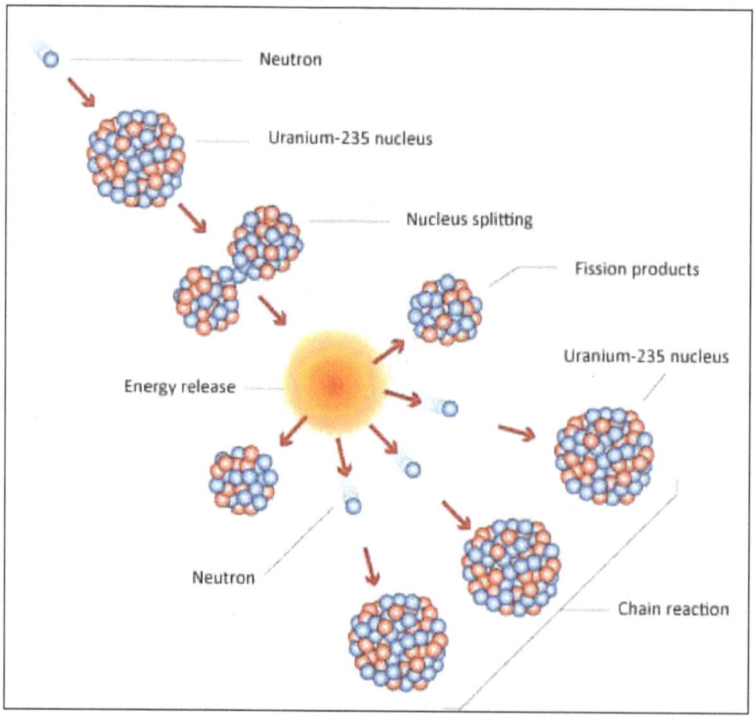

Figure: A chain reaction of uranium fission is a powerful source of energy.

Radioactive Disintegration Law

The rate of radioactive disintegration is always directly proportional to the number of atoms present in the sample at a given point of time. Radioactive disintegration can be expressed mathematically as follows, since the radioactive decay increases exponentially the mathematical expression imprints the same: Let the number of radioactive atoms present in the sample originally be N_o and let the number of radioactive atoms at the given point in time *t* be N*N*. Then,

$$N = N_0 \cdot e^{-\lambda t}$$

Where λ is the decay constant, or the disintegration constant, which can be determined as follows:

$$\frac{N_0 \cdot e^{-\lambda t}}{N} = 1$$

$$\ln\left(\frac{N_0}{N}\right) + \ln\left(e^{-\lambda t}\right) = \ln(1)$$

$$\ln\left(\frac{N_0}{N}\right) - \lambda t = 0$$

$$\Rightarrow \ln\left(\frac{N_0}{N}\right) = \lambda t$$

Radioactive Half-life

The half-life of a radioactive element is the time that it takes for half the nuclei in the sample to decay in a first-order reaction. The half-life of a radioisotope can be fractions of a second or millions of years, depending on the element.

Half-life

Half-life ($t_{1/2}$) is defined as the time taken for half of the original number of atoms in a radioactive sample to disintegrate. The half-life remains constant. Even if the sample has undergone one half-life, the time period for the next half-life remains unchanged.

Using the equation introduced above, $N = N_o . e^{-\lambda t}$ And the definition of half-life, when $t = t_{1/2}, N = \frac{N_0}{2}$. So,

$$\frac{N_0}{2} = N_0 e^{\lambda t_{1/2}}$$

$$e^{\lambda t_{1/2}} = 2$$

$$\lambda t_{1/2} = \ln 2$$

$$\Rightarrow t_{1/2} = \frac{0.693}{\lambda}$$

The average lifetime of a radioactive material is the inverse of the decay constant.

$$\text{Average lifetime} = \frac{1}{\lambda}$$

The Manhattan Project

During World War II, both Germany and the United States began experimenting with the use of nuclear reactions to build weapons, with the United States succeeding first and dropping two atomic bombs on Japanese cities in August 1945. The Manhattan Project, which led to the development of the U.S. Department of Energy and the national laboratories, was a secret project that connects far-flung facilities throughout the country, including Oak Ridge, TN; Chicago, IL; Los Alamos, NM; and Hanford, WA. These facilities were often in rural areas both due to the secretive nature of the work and due to the space and machinery needed to process the elements for the bomb's cores and to test the weapons.

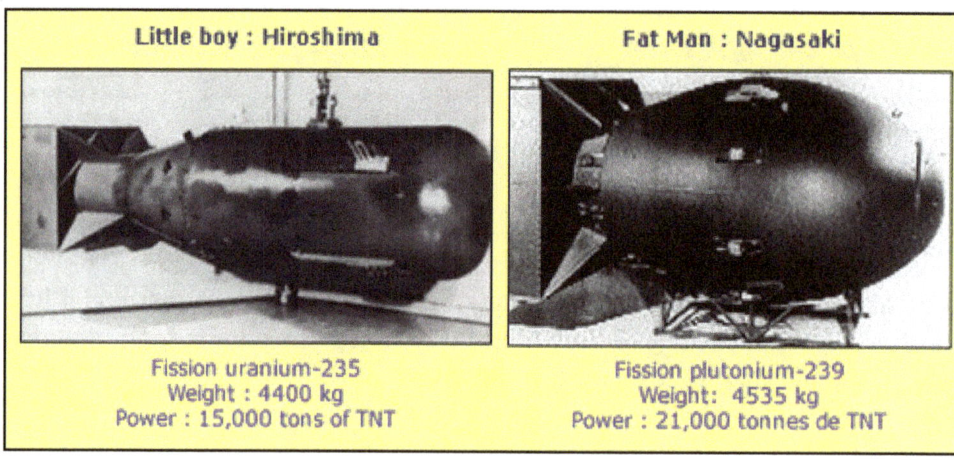

The destructive power of atomic weapons comes from the energy produced by splitting the nuclei of the elements in the bombs' core. The U.S. developed two types of atomic bombs during the World War II. The first, which was nicknamed Little Boy, was dropped on the Japanese city of Hiroshima was a gun-type weapon with a uranium core. The second weapon, dropped on Nagasaki, was called Fat Man and was an implosion-type device with a plutonium core.

The exact designs of nuclear weapons are not publicly available, but the basic scientific concepts are discussed below. The isotopes uranium-235 and plutonium-239 were selected for their ability to easily undergo fission. When a neutron strikes the nucleus of an atom belonging to one of these species, it splits the nucleus into fragments and releases energy. The fission process becomes self-sustaining as neutrons produced by the splitting of atom strike nearby nuclei and produce more fission. This is known as a chain reaction and it ultimately leads to an atomic explosion.

Little Boy was powered by the uranium isotope U-235, which required an extensive extraction and enrichment process. Most uranium found naturally in the world exists as U-238, with less than 1% naturally existing as the U-235 isotope. When a neutron bombards U-238, the isotope often captures the neutron to become U-239, failing to fission or instigate the chain reaction necessary to detonate a bomb. The first challenge of the Manhattan Project was to determine the most efficient way to separate and purify U-235. Once enough U-235 was obtained to power the bomb, Little Boy was constructed using a gun-type design that fired one portion of U-235 at another inside the bomb. The combined masses reached the critical point to set off a fission chain reaction.

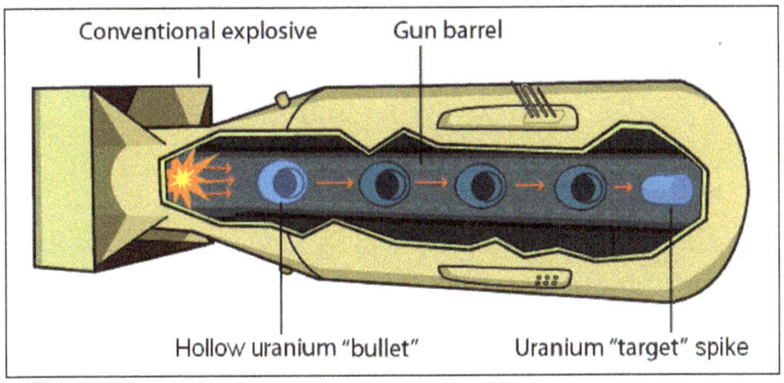

Figure: Little Boy used a gun-type design.

Fat Man could not use the same gun-type design that allowed Little Boy to explode effectively—the form of plutonium collected from the nuclear reactors at Hanford, WA for the bomb would not allow for this strategy due to the presence of traces of the isotope Pu-240. Plutonium-240's higher fission rate would cause the atoms to undergo spontaneous fission before the gun-type design could bring the two masses of plutonium together, which would lower the energy involved in the actual detonation of the bomb.

The new design used conventional explosives around a central plutonium mass to consolidate the plutonium, increasing the pressure and density. An increased density allowed the plutonium to reach its critical mass. To detonate the bomb, the explosives were ignited, releasing a shock wave that compressed the inner plutonium and led to its explosion.

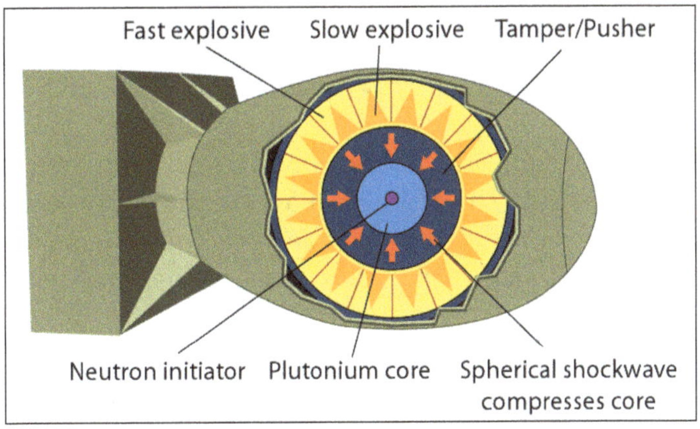

Figure: Fat Man, an implosion-type bomb.

Alpha Decay

Alpha decay or α-decay is a type of radioactive decay in which the atomic nucleus emits an alpha particle thereby transforming or decaying into a new atomic nucleus. Here the atomic mass number of the newly formed atom will be reduced by four and the atomic number will be reduced by two. The emitted alpha particle is also known as a helium nucleus. The mass of the alpha particles is relatively large and has a positive charge.

Ernest Rutherford distinguished alpha decay from other forms of radiation by studying the deflection of the radiation through a magnetic field. The deflection of alpha decay would be a positive charge as the particles have a +2e charge.

Equation for Alpha Decay

Following is the general alpha decay equation:

$$^{A}_{Z}X \rightarrow ^{A-4}_{Z-4}Y + ^{4}_{2}He$$

where,

- $^{A}_{Z}X$ is the parent nucleus.
- A is the total no.of nucleons.
- Z is the total no.of protons.
- $^{A-4}_{Z-4}Y$ is the daughter nucleus.
- $^{4}_{2}He$ is the released alpha particle.

Alpha decay is a type of radioactive decay in which a particle with two neutrons and two protons is ejected from the nucleus of a radioactive atom.

Alpha Decay Symbol

Alpha decay is denoted by α or He2+ or α2+.

What Happens in Alpha Decay?

Alpha decay occurs in very heavy elements like uranium, thorium, and radium. They are called parent nucleus and they are basically unstable. Because the nuclei of these atoms have a lot more neutrons in their nuclei than protons, that is, they have too large a proton to neutron ratio, which makes these elements neutron-rich. This richness makes alpha decay possible. Thus, emitting its two protons and two neutrons in the form of an alpha particle and a forming of new daughter nucleus and attains a very stable configuration. Alpha decay can be described like this:

- The nucleus of these nuclei (parent nucleus) rich atoms splits into two parts.
- The alpha particle goes zooming off into space.
- The nucleus left behind (daughter nucleus) has its atomic number reduced by 2 and its mass number reduced by 4.

Alpha Decay Formula

In general, the following is the way to write the formula for alpha decay:

$$^{A}_{Z}X \rightarrow ^{A-4}_{Z-4}X' + ^{4}_{2}\alpha$$

Where,

- ${}^{A}_{Z}X$ is the parent nucleus.
- ${}^{A-4}_{Z-4}X'$ is the daughter nucleus.
- ${}^{4}_{2}\alpha$ is the emitted alpha particle.

Alpha Decay Equation

In nuclear physics, Alpha decay Equation for thorium-232 can be written as,

$${}^{232}_{90}Th \rightarrow {}^{228}_{88}Ra + {}^{4}_{2}He$$

Where,

- ${}^{232}_{90}Th$ is the unstable Thorium-232 parent nucleus.
- ${}^{228}_{88}Ra$ is the stable Radium-228 daughter nucleus.
- ${}^{4}_{2}He$ is the emitted alpha particle.

Alpha Decay Example

Ample of Alpha decay example can be given. Some of them are,

$${}^{237}_{93}Np \rightarrow {}^{233}_{91}Pa + {}^{4}_{2}He$$

$${}^{175}_{78}Pt \rightarrow {}^{171}_{76}Os + {}^{4}_{2}He$$

$${}^{149}_{64}Gd \rightarrow {}^{145}_{62}Sm + {}^{4}_{2}He$$

Alpha Decay of Uranium 238

Alpha decay of Uranium 238 heavy nucleus into Thorium 234 with the emission of an alpha particle is given as:

$${}^{238}_{92}Ur \rightarrow {}^{234}_{909}Th + {}^{4}_{2}He$$

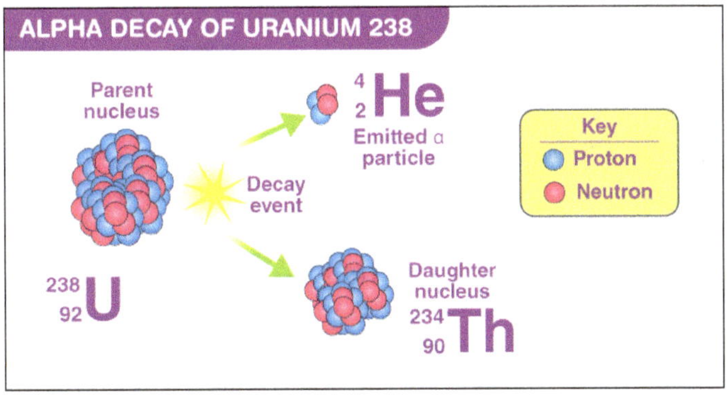

Gamow Theory of Alpha Decay

The Geiger–Nuttall law or Geiger–Nuttall rule relates to the decay constant of a radioactive isotope with the energy of the alpha particles emitted. This relation also states that half-lives are exponentially dependent on decay energy, so that very large changes in half-life make comparatively small differences in decay energy, and thus alpha particle energy.

As per this rule, short-lived isotopes emit more energetic alpha particles than long-lived ones. This law was stated by Hans Geiger and John Mitchell Nuttall in the year 1911; hence the name was dedicated to these physicists.

Beta Decay

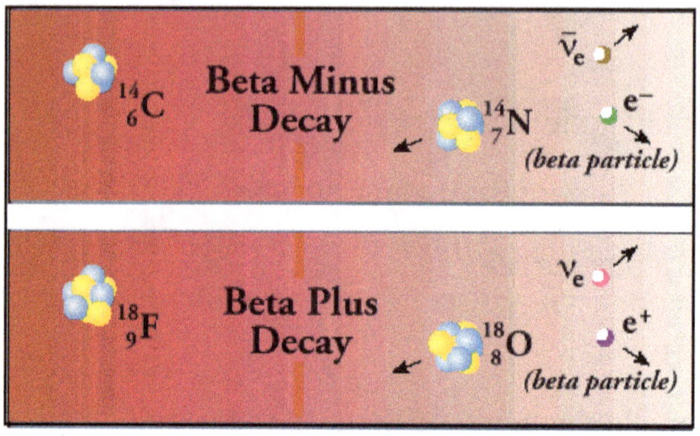

Figure: Beta decays.

Beta particles are electrons or positrons (electrons with positive electric charge, or antielectrons). Beta decay occurs when, in a nucleus with too many protons or too many neutrons, one of the protons or neutrons is transformed into the other. In beta minus decay, a neutron decays into a proton, an electron, and an antineutrino: $n \varnothing p + e - +$. Beta plus decay, a proton decays into a neutron, a positron, and a neutrino: $p \varnothing n + e + + \nu$. Both reactions occur because in different regions of the Chart of the Nuclides, one or the other will move the product closer to the region of stability. These particular reactions take place because conservation laws are obeyed. Electric charge conservation requires that if an electrically neutral neutron becomes a positively charged proton, an electrically negative particle (in this case, an electron) must also be produced. Similarly, conservation of lepton number requires that if a neutron (lepton number = 0) decays into a proton (lepton number = 0) and an electron (lepton number = 1), a particle with a lepton number of -1 (in this case an antineutrino) must also be produced. The leptons emitted in beta decay did not exist in the nucleus before the decay–they are created at the instant of the decay.

To the best of our knowledge, an isolated proton, a hydrogen nucleus with or without an electron, does not decay. However within a nucleus, the beta decay process can change a proton to a neutron. An isolated neutron is unstable and will decay with a half-life of 10.5 minutes. A neutron in a nucleus will decay if a more stable nucleus results; the half-life of the decay depends on the isotope. If it leads to a more stable nucleus, a proton in a nucleus may capture an electron from the atom (electron capture), and change into a neutron and a neutrino.

Proton decay, neutron decay, and electron capture are three ways in which protons can be changed into neutrons or vice-versa; in each decay there is a change in the atomic number, so that the parent and daughter atoms are different elements. In all three processes, the number A of nucleons remains the same, while both proton number, Z, and neutron number, N, increase or decrease by 1.

In beta decay the change in binding energy appears as the mass energy and kinetic energy of the beta particle, the energy of the neutrino, and the kinetic energy of the recoiling daughter nucleus. The energy of an emitted beta particle from a particular decay can take on a range of values because the energy can be shared in many ways among the three particles while still obeying energy and momentum conservation.

Gamma Decay

Gamma decay is one type of radioactive decay that a nucleus can undergo. What separates this type of decay process from alpha or beta decay is that no particles are ejected from the nucleus when it undergoes this type of decay. Instead, a high energy form of electromagnetic radiation - a gamma ray photon - is released. Gamma rays are simply photons that have extremely high energies which are highly ionizing. As well, gamma radiation is unique in the sense that undergoing gamma decay does not change the structure or composition of the atom. Instead, it only changes the energy of the atom since the gamma ray carries no charge nor does it have an associated mass.

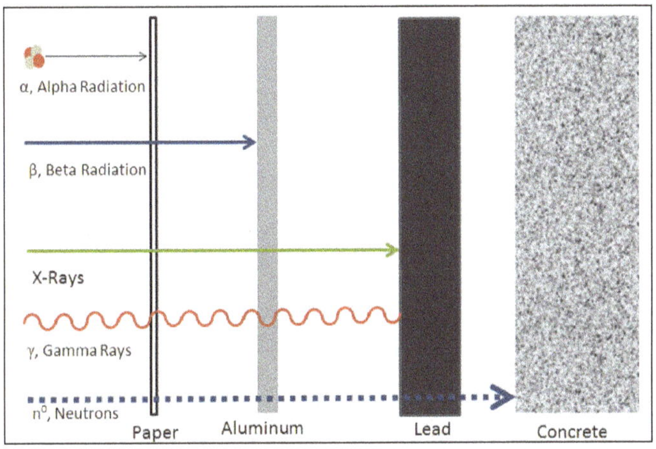

Figure: Different penetration levels of different products of decay, with gamma being one of the most highly penetrating.

In order for a nucleus to undergo gamma decay, it must be in some sort of excited energetic state. Experiments have shown that protons and neutrons are located in discrete energy states within the nucleus, not too different from the excited states that electrons can occupy in atoms. Thus if a proton or a neutron inside of the nucleus jumps up to an excited state - generally following an alpha or beta decay - the new daughter nucleus must somehow release energy to allow the proton or neutron to relax back down to ground state. When the nucleon makes this transition from a high to a low energy state, a gamma photon is emitted. The general equation that represents this process is:

$$A^* \rightarrow A + \gamma$$

where:

- A^* is the excited atom.
- A is a relaxed state of the initial excited atom.

- γ is the released gamma ray photon.

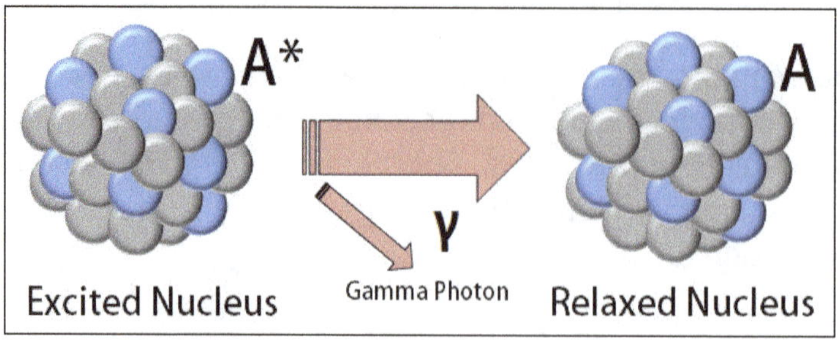

Figure: Diagram showing gamma decay of a nucleus.

Knowing that an atom undergoes gamma radiation is important, but it is also possible to determine the frequency of the released gamma radiation if the initial and final states of the nucleon inside the nucleus are known. The equation representing the frequency of the gamma radiation is:

$$E_i - E_f = hf$$

where:

- E_i is the initial, higher energy state of the nucleon.
- E_f is the final, lower energy state of the nucleon.
- h is Planck's constant.
- f is the frequency of the emitted radiation.

In addition to radioactive nuclei, there are many objects in space that emit high levels of gamma radiation.

Applications and Importance

Gamma rays can at times be harmful due to the fact that they are generally very high energy and therefore penetrate matter very easily. Since it penetrates so easily, it is some of the most useful radiation for medical purposes.

Some of the most widely used gamma emitters are cobalt-60, cesium-137, and technetium-99m. Cesium is used widely in radiotherapy - the treatment of cancer using gamma rays - as well as being used to measure soil density at construction sites and to investigate the subterranean layers of the Earth in oil wells. Cobalt is used to sterilize medical equipment and irradiate food, killing bacteria and pasteurizing the food. Technetium-99m (which has a shorter half-life than technetium-99) is the most widely used for diagnostic medical tests to investigate the brain, bone, and internal organs. As well, exposure to gamma radiation can improve the durability of wood and plastics, and is thus used to toughen flooring in high-traffic areas.

In addition, uranium-238 and uranium-235 - used in fuel for nuclear power plants - undergo both alpha and gamma decays when used. Immediately following the fission process, gamma rays are released, resulting in high levels of radiation present around the reactor. However, safety precautions are in line to ensure that workers do not get close enough to this radioactive area to be harmed.

Conservation Laws in Nuclear Reactions

In analyzing nuclear reactions, we apply the many conservation laws. Nuclear reactions are subject to classical conservation laws for charge, momentum, angular momentum, and energy (including rest energies). Additional conservation laws, not anticipated by classical physics, are are electric charge, lepton number and baryon number. Certain of these laws are obeyed under all circumstances, others are not. We have accepted conservation of energy and momentum. In all the examples given we assume that the number of protons and the number of neutrons is separately conserved. We shall find circumstances and conditions in which this rule is not true. Where we are considering non-relativistic nuclear reactions, it is essentially true. However, where we are considering relativistic nuclear energies or those involving the weak interactions, we shall find that these principles must be extended.

Some conservation principles have arisen from theoretical considerations, others are just empirical relationships. Notwithstanding, any reaction not expressly forbidden by the conservation laws will generally occur, if perhaps at a slow rate. This expectation is based on quantum mechanics. Unless the barrier between the initial and final states is infinitely high, there is always a non-zero probability that a system will make the transition between them.

- Conservation of nucleons: The total number of nucleons before and after a reaction is the same.
- Conservation of charge: The sum of the charges on all the particles before and after a reaction are the same.
- Conservation of momentum: The total momentum of the interacting particles before and after a reaction is the same.
- Conservation of energy: Energy, including rest mass energy, is conserved in nuclear reactions.

Although the number of possible nuclear reactions is enormous, nuclear reactions can be sorted by types. Most of nuclear reactions are accompanied by gamma emission. Some examples are:

- Elastic scattering: Occurs, when no energy is transferred between the target nucleus and the incident particle.

  208Pb (n, n) 208Pb

- Inelastic scattering: Occurs, when energy is transferred. The difference of kinetic energies is saved in excited nuclide.

  40Ca (α, α') 40mCa

- Capture reactions: Both charged and neutral particles can be captured by nuclei. This is accompanied by the emission of γ-rays. Neutron capture reaction produces radioactive nuclides (induced radioactivity).

  238U (n, γ) 239U

- Transfer Reactions: The absorption of a particle accompanied by the emission of one or more particles is called the transfer reaction.

  4He (α, p) 7Li

- Fission reactions: Nuclear fission is a nuclear reaction in which the nucleus of an atom splits into smaller parts (lighter nuclei). The fission process often produces free neutron-sand photons (in the form of gamma rays), and releases a large amount of energy.

  235U (n, 3 n) fission products

- Fusion reactions: Occur when, two or more atomic nuclei collide at a very high speed and join to form a new type of atomic nucleus.The fusion reaction of deuterium and tritium is particularly interesting because of its potential of providing energy for the future.

  3T (d, n) 4He

- Spallation reactions: Occur, when a nucleus is hit by a particle with sufficient energy and momentum to knock out several small fragments or, smash it into many fragments.

- Nuclear decay (Radioactive decay): Occurs when an unstable atom loses energy by emitting ionizing radiation. Radioactive decay is a random process at the level of single atoms, in that, according to quantum theory, it is impossible to predict when a particular atom will decay. There are many types of radioactive decay:

  - Alpha radioactivity: Alpha particles consist of two protons and two neutrons bound together into a particle identical to a helium nucleus. Because of its very large mass (more than 7000 times the mass of the beta particle) and its charge, it heavy ionizes material and has a very short range.

  - Beta radioactivity: Beta particles are high-energy, high-speed electrons or positrons emitted by certain types of radioactive nuclei such as potassium-40. The beta particles have greater range of penetration than alpha particles, but still much less than gamma rays.The beta particles emitted are a form of ionizing radiation also known as beta rays. The production of beta particles is termed beta decay.

  - Gamma radioactivity: Gamma rays are electromagnetic radiation of an very high frequency and are therefore high energy photons. They are produced by the decay of nuclei as they transition from a high energy state to a lower state known as gamma decay. Most of nuclear reactions are accompanied by gamma emission.

- Neutron emission: Neutron emission is a type of radioactive decay of nuclei containing excess neutrons (especially fission products), in which a neutron is simply ejected from the nucleus. This type of radiation plays key role in nuclear reactor control, because these neutrons are delayed neutrons.

Conservation of Energy in Nuclear Reactions

In analyzing nuclear reactions, we have to apply the general law of conservation of mass-energy. According to this law mass and energy are equivalent and convertible one into the other. It is one of the striking results of Einstein's theory of relativity. This equivalence of the mass and energy is described by Einstein's famous formula $E = mc^2$.

Generally, in both chemical and nuclear reactions, some conversion between rest mass and energy occurs, so that the products generally have smaller or greater mass than the reactants. In general, the total (relativistic) energy must be conserved. The "missing" rest mass must therefore reappear as kinetic energy released in the reaction. The difference is a measure of the nuclear binding energy which holds the nucleus together. The nuclear binding energies are enormous; they are of the order of a million times greater than the electron binding energies of atoms.

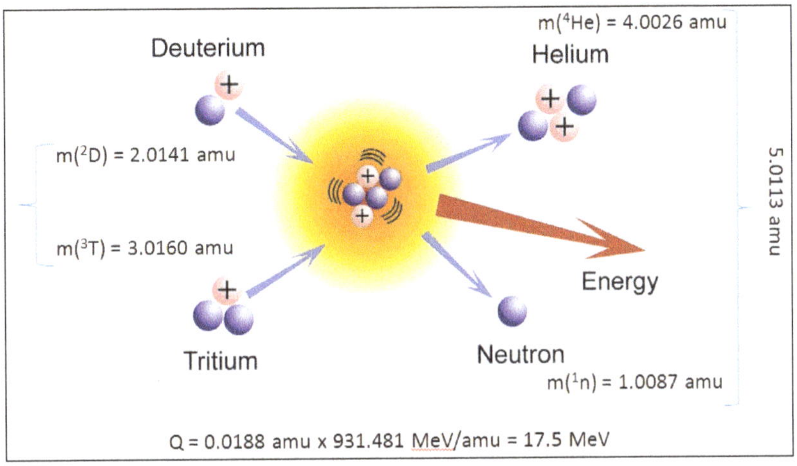

Figure: Q-value of DT fusion reaction.

The energetics of nuclear reactions is determined by the Q-value of that reaction. The Q-value of the reaction is defined as the difference between the sum of the masses of the initial reactants and the sum of the masses of the final products, in energy units (usually in MeV).

Consider a typical reaction, in which the projectile a and the target A gives place to two products, B and b. This can also be expressed in the notation that we used so far, $a + A \rightarrow B + b$, or even in a more compact notation, A(a,b)B.

The Q-value of this reaction is given by:

$$Q = [m_a + m_A - (m_b + m_B)]c^2$$

which is the same as the excess kinetic energy of the final products:

$$Q = T_{final} - T_{initial}$$

$$= T_b + T_B - (T_a + T_A)$$

For reactions in which there is an increase in the kinetic energy of the products Q is positive. The positive Q reactions are said to be exothermic (or exergic). There is a net release of energy, since the kinetic energy of the final state is greater than the kinetic energy of the initial state.

For reactions in which there is a decrease in the kinetic energy of the products Q is negative. The negative Q reactions are said to be endothermic (or endoergic) and they require a net energy input.

Example Exothermic Reaction - DT Fusion

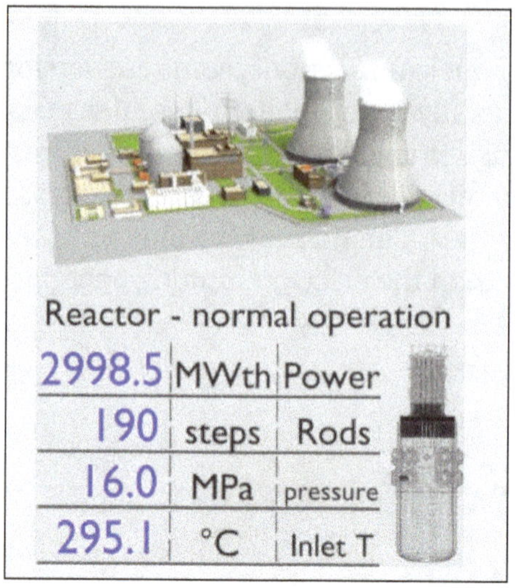

Example Endothermic Reaction - Photoneutrons

In nuclear reactors the gamma radiation plays a significant role also in reactor kinetics and in a subcriticality control. Especially in nuclear reactors with D2O moderator (CANDU reactors) or with Be reflectors (some experimental reactors). Neutrons can be produced also in (γ, n) reactions and therefore they are usually referred to as photoneutrons.

A high energy photon (gamma ray) can under certain conditions eject a neutron from a nucleus. It occurs when its energy exceeds the binding energy of the neutron in the nucleus. Most nuclei have binding energies in excess of 6 MeV, which is above the energy of most gamma rays from fission. On the other hand there are few nuclei with sufficiently low binding energy to be of practical interest. These are: ^{2}D, ^{9}Be, ^{6}Li, ^{7}Li and ^{13}C. As can be seen from the table the lowest thresholds have ^{9}Be with 1.666 MeV and ^{2}D with 2.226 MeV.

Nuclide	Threshold(MeV)	Reaction
^{2}D	2.225	$^{2}H(\gamma,n)^{1}H$
^{6}Li	3.697	$^{6}Li(\gamma,n+p)^{4}He$
^{6}Li	5.67	$^{6}Li(\gamma,n)^{5}Li$
^{7}Li	7.251	$^{7}Li(\gamma,n)^{6}Li$
^{9}Be	1.667	$^{9}Be(\gamma,n)^{8}Be$
^{13}C	4.9	$^{13}C(\gamma,n)^{12}C$

In case of deuterium, neutrons can be produced by the interaction of gamma rays (with a minimum energy of 2.22 MeV) with deuterium:

$$^{2}_{1}D+\gamma \rightarrow ^{1}_{1}H+n$$

The reaction Q-value is calculated below:

The atom masses of the reactant and products are:

$m(^2D) = 2.01363$ amu

$m(^1n) = 1.00866$ amu

$m(^1H) = 1.00728$ amu

Using the mass-energy equivalence, we get the Q-value of this reaction as:

$$Q = \{2.01363\ [amu] - (1.00866+1.00728)\ [amu]\} \times 931.481\ [MeV/amu]$$

$$= -0.00231 \times 931.481 = -2.15\ MeV$$

The energy released in a nuclear reaction can appear mainly in one of three ways:

- Kinetic energy of the products.
- Emission of gamma rays: Gamma rays are emitted by unstable nuclei in their transition from a high energy state to a lower state known as gamma decay.
- Metastable state: Some energy may remain in the nucleus, as a metastable energy level.

A small amount of energy may also emerge in the form of X-rays. Generally, products of nuclear reactions may have different atomic numbers, and thus the configuration of their electron shells is different in comparison with reactants. As the electrons rearrange themselves and drop to lower energy levels, internal transition X-rays (X-rays with precisely defined emission lines) may be emitted.

Conservation of Momentum and Energy in Collisions

The use of the conservation laws for momentum and energy is very important also in particle collisions. This is a very powerful rule because it can allow us to determine the results of a collision without knowing the details of the collision. The law of conservation of momentum states that in the collision of two objects such as billiard balls, the total momentum is conserved. The assumption of conservation of momentum as well as the conservation of kinetic energy makes possible the calculation of the final velocities in two-body collisions. At this point we have to distinguish between two types of collisions:

- Elastic collisions,
- Inelastic collisions.

Elastic Collisions

A perfectly elastic collision is defined as one in which there is no net conversion of kinetic energy into other forms (such as heat or noise). For the brief moment during which the two objects are in contact, some (or all) of the energy is stored momentarily in the form of elastic potential energy. But if we compare the total kinetic energy just before the collision with the total kinetic energy just after the collision, and they are found to be the same, then we say that the total kinetic energy is conserved.

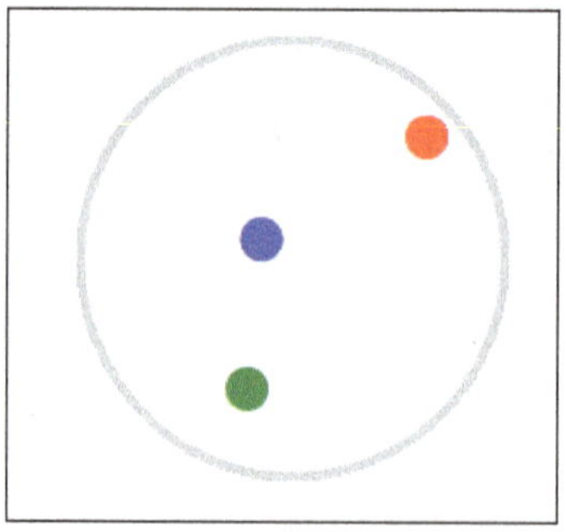

- Some large-scale interactions like the slingshot type gravitational interactions (also known as a planetary swing-by or a gravity-assist manoeuvre) between satellites and planets are perfectly elastic.
- Collisions between very hard spheres may be nearly elastic, so it is useful to calculate the limiting case of an elastic collision.
- Collisions in ideal gases approach perfectly elastic collisions, as do scattering interactions of sub-atomic particles which are deflected by the electromagnetic force.
- Rutherford scattering is the elastic scattering of charged particles also by the electromagnetic force.
- A neutron-nucleus scattering reaction may be also elastic, but in this case the neutron is deflected by the strong nuclear force.

Equations of Conservation of Momentum and Energy

Let us assume the one dimensional elastic collision of two objects, the object A and the object B. These two objects are moving with velocities vA and vB along the x axis before the collision. After the collision, their velocities are v'A and v'B. The conservation of the total momentum demands that the total momentum before the collision is the same as the total momentum after the collision. Likewise, the conservation of the total kinetic energy, which demands that the total kinetic energy of both objects before the collision is the same as the total kinetic energy after the collision. Both law may be expressed in equations as:

Total momentum before= Total momentum after:

$$m_A v_A + m_B v_B = m_A v'_A + m_B v'_B$$

Total kinetic energy before= Total kinetic energy after:

$$\frac{1}{2} m_A v_A^2 + \frac{1}{2} m_B v_B^2 = \frac{1}{2} m_A v'^2_A + m_B v'^2_B$$

We have two equations, so that we can solve for two unknowns. To solve these simultaneous equations for v^1_A and v^1_B, we first rewrite the both equations as:

$$m_A(v_A - v'_A) = m_B(v'_B - v_B)$$

$$m_A(v_A^2 - v'^2_A) = m_B(v'^2_B - v_B^2)$$

When we use a²-b²= (a-b) (a=b):

$$m_A(v_A - v'_A)(v_A - v'_A) = m_B(v'_B - v_B)(v'_B - v_B)$$

When we divide this equation for kinetic energies by equation for momentum, we obtain:

$$v_A + v'_A = v'_B - v_B$$

The relative speed of the two objects after the collision has the same magnitude (but opposite direction) as before the collision, no matter what the masses are. It is known the fission neutrons are of importance in any chain-reacting system. All neutrons produced by fission are born as fast neutrons with high kinetic energy. Before such neutrons can efficiently cause additional fissions, they must be slowed down by collisions with nuclei in the moderator of the reactor. The probability of the fission U-235 becomes very large at the thermal energies of slow neutrons. This fact implies increase of multiplication factor of the reactor (i.e. lower fuel enrichment is needed to sustain chain reaction).

The neutrons released during fission with an average energy of 2 MeV in a reactor on average undergo a number of collisions (elastic or inelastic) before they are absorbed. During the scattering reaction, a fraction of the neutron's kinetic energy is transferred to the nucleus. Using the laws of conservation of momentum and energy and the analogy of collisions of billiard balls for elastic scattering, it is possible to derive the following equation for the mass of target or moderator nucleus (M), energy of incident neutron (Ei) and the energy of scattered neutron (E_s).

$$E_s = \left(\frac{M - m_n}{M + m_n}\right)^2 E_i \quad E_s = \left(\frac{A-1}{A+1}\right)^2 E_i$$

where A is the atomic mass number.In case of the hydrogen (A = 1) as the target nucleus, the incident neutron can be completely stopped. But this works when the direction of the neutron is completely reversed (i.e. scattered at 180°). In reality, the direction of scattering ranges from 0 to 180 ° and the energy transferred also ranges from 0% to maximum. Therefore, the average energy of scattered neutron is taken as the average of energies with scattering angle 0 and 180°.

Moreover, it is useful to work with logarithmic quantities and therefore one defines the logarithmic energy decrement per collision (ξ) as a key material constant describing energy transfers during a neutron slowing down. ξ is not dependent on energy, only on A and is defined as follows:

$$\xi = \overline{\ln E' - \ln E} = \int_{\alpha E'}^{E'} \ln \frac{E'}{E} p(E' \to E) dE = 1 + \frac{\alpha}{1-\alpha} \ln \alpha$$

Where,

$$\left(\frac{A-1}{A+1}\right)^2 = \alpha$$

For heavy target nuclei, ξ may be approximated by following formula:

$$\xi \simeq \frac{2}{A+2/3} \quad \text{for } A>10$$

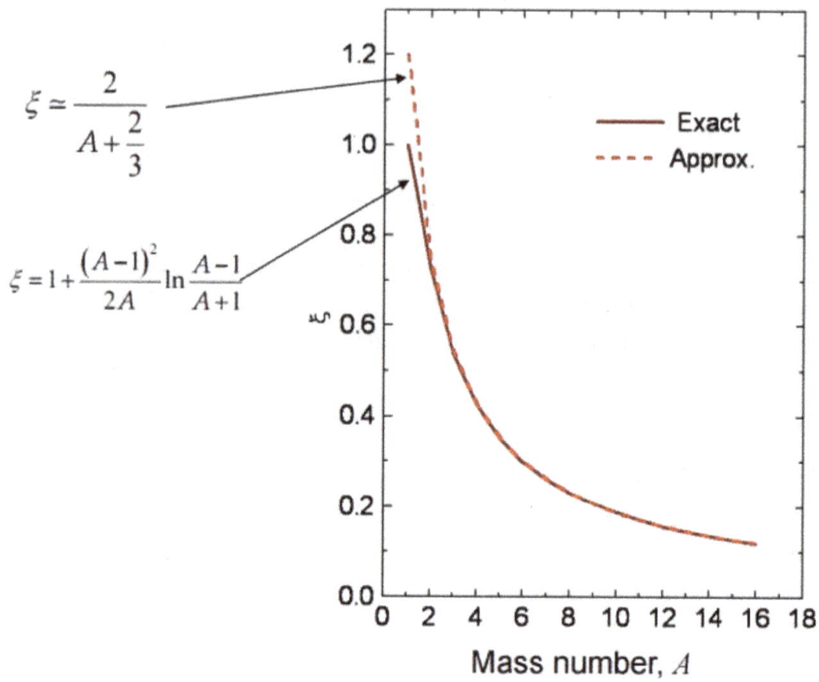

From these equations it is easy to determine the number of collisions required to slow down a neutron from, for example from 2 MeV to 1 eV.

Example: Determine the number of collisions required for thermalization for the 2 MeV neutron in the carbon.

$\xi_{CARBON} = 0.158$

$N(2MeV \rightarrow 1eV) = \ln 2{\cdot}10^6/\xi = 14.5/0.158 = 92.$

Table of average logarithmic energy decrement for some elements.

Elements	A	ξ	Collisions 2 mev→1ev
H	1	1	15
D	2	0.750	20
H_2O	-	0.920	16
D_2O	-	0.509	29
He	4	0.425	34
Be	9	0.207	70

C	12	0.158	92
O	16	0.120	121
Na	23	0.084	172
Fe	56	0.035	414
^{238}U	238	0.008	1812

For a mixture of isotopes:

$$\bar{\xi} = \frac{Sum(\xi_i.\Sigma_{Si})}{Sum(\Sigma_{Si})}$$

Example Elastic Nuclear Collision

A neutron (n) of mass 1.01 u traveling with a speed of 3.60×10^4m/s interacts with a carbon (C) nucleus (m_C = 12.00 u) initially at rest in an elastic head-on collision. What are the velocities of the neutron and carbon nucleus after the collision?

Solution: This is an elastic head-on collision of two objects with unequal masses. We have to use the conservation laws of momentum and of kinetic energy, and apply them to our system of two particles.

- Conservation of Momentum.

  Total P before= Total P after

  $$m_n v_n + 0 = m_n v'_n = m_c v'_c$$

- Conservation of kinetic energy.

  Total K before= Total K after

  $$\frac{1}{2} m_n v_n^2 + 0 = \frac{1}{2} m_n v'^2_n + \frac{1}{2} m_c v'^2_c$$

We can solve this system of equation or we can use the equation. This equation stated that the relative speed of the two objects after the collision has the same magnitude (but opposite direction) as before the collision, no matter what the masses are.

$$v_A + v'_A = v_B + v'_B$$

In our case:

$$v_n - 0 = v_c^{'} - v_n^{'}$$

Thus,

$$v_n^{'} = v_c^{'} - v_n$$

Substituting this into our momentum equation, we get:

$$m_n v_n = m_n v'_c - m_n v_n + m_n v'_c$$

Solving this equation for both velocities, we obtain:

$$v'_c = \frac{2m_n v_n}{m_n + m_c} = \frac{2x1.01}{13.01} x3.6x10^4 = 0.56x10^4 m/s$$

$$v_n' = v_c' - v_n = 0.56x10^4 - x3.6x10^4 m/s$$

The minus sign for v' tell us that the neutron scatters back of the carbon nucleus, because the carbon nucleus is significantly heavier. On the other hand its speed is less than its initial speed. This process is known as the neutron moderation and it significantly depends on the mass of moderator nuclei.

References

- Nuclear-fission, boundless-chemistry: courses.lumenlearning.com, Retrieved 11, July 2020
- Elastic-scattering, earth-and-planetary-sciences: sciencedirect.com, Retrieved 30, June 2020
- Neutron-inelastic-scattering, neutron-nuclear-reactions, nuclear-engineering-fundamentals, nuclear-power, reactor-physics: nuclear-power.net, Retrieved 24, March 2020
- Alpha-decay, physics: byjus.com, Retrieved 09, February 2020
- conservation-laws-in-nuclear-reactions, laws-of-conservation: nuclear-power.net, Retrieved 09, May 2020

Chapter 6

Heat Transfer Processes

Heat flow process deals with the generation, transfer, use and cooling of thermal energy within a system. Heat generation, neutron flux, steam generator, heat transport, emergency core cooling system, moderator and moderator system are some of the aspects that fall under its domain. This chapter has been carefully written to provide an easy understanding of all these topics with relation to their implementation in nuclear reactors.

Heat Generation

The principle modes of heat generation in a nuclear reactor are the reactions (nuclear reactions) that neutrons undergo with nuclei of various materials used in the reactor.

Nuclear fission is exothermic with the energy released due to fission attributed to splitting of heavy nuclei into fragments. The energy released is proportional to the difference in mass of the reactants (neutron & nucleus) and that of products (fission fragments). On an average, 207 MeV (1 MeV = 1.61×10^{-13} J) of energy is released per fission event in U-235, of which around 200 MeV can be recovered as heat. This energy is sum of the kinetic energies & decay energy of the fission fragments, kinetic energy of new neutrons and the energy of gamma radiation. This energy can be further classified into energy release due to fission and neutron capture.

The recoverable energy released due to fission (200 MeV) can be further classified into instantaneous energy and delayed energy. The components of instantaneous energy release are the kinetic energy of fission fragments (168 MeV), kinetic energy of new neutrons (5 MeV) and gamma radiation (7 MeV). The contributions for the delayed energy come from β-decay (8 MeV) and γ-decay (7 MeV) of fission fragments.

The energy released due to neutron capture is the result of non-fission reactions between excess neutrons (unutilized for chain reaction) and their β-decay and γ-decay (5 MeV).

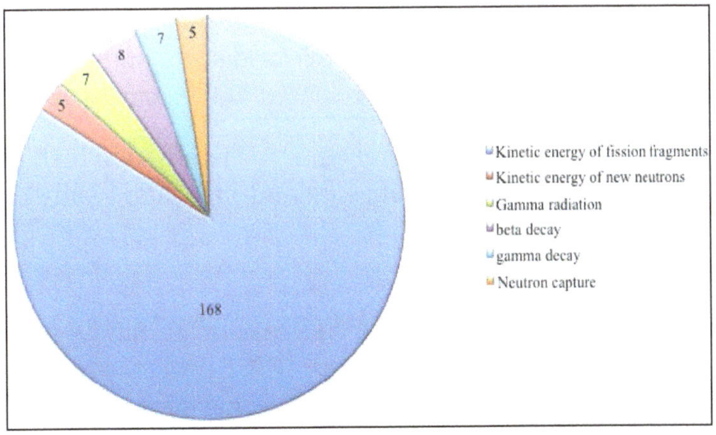

Figure illustrates the energy contribution from a nuclear reactor using U-235 as the fissile isotope. The contributions of various nuclear reactions towards the recoverable energy are also shown.

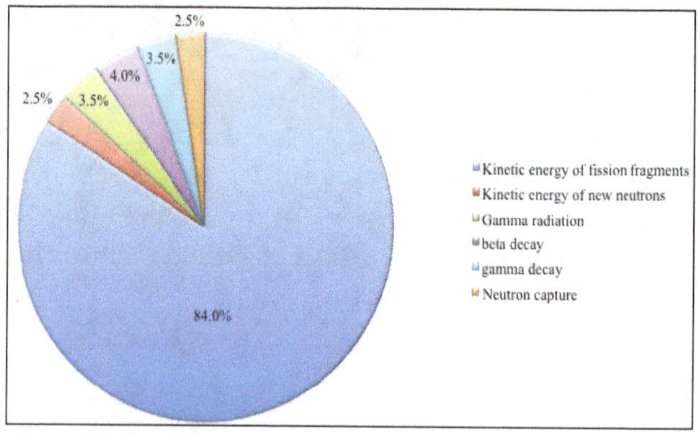

Contribution of different nuclear reactions towards recoverable heat energy: (a) energy in MeV; (b) energy contribution in % of total recoverable energy.

Now let's discuss the distribution of this energy among various components of the reactor.

The fission fragments have very short range (< 0.25 mm). The β-radiation too has short range (< 1 mm). Hence their energy is released within the fuel element itself.

The energy released during thermalization of high-energy neutrons rests in the moderator. The energy of the gamma radiation is released in the fuel as well as in structural elements. Table shows the distribution of energy in various components of the reactor.

Table: Distribution of fission energy in various components of the thermal reactor.

Energy Source	Location of its release
Kinetic energy of fission fragments	Fuel
Kinetic energy of new neutrons	Moderator
γ-radiation (instantaneous)	Fuel and structural components
γ-radiation (delayed)	Fuel and structural components
β-decay of fission products (delayed)	Fuel
Neutron capture (delayed)	Fuel and structural components

For a light-water reactor, 92 % of the total energy released stays in the fuel while about 3 % is released in the moderator. The remaining (5 %) is released in the structural elements.

In a CANDU-type reactor, about 94 % of the total energy is released in fuel, 5% is released in moderator, while the rest is released in pressure tubes, calandria, coolant and shielding.

Reactor Power

Let us look at the factors that influence the power of a reactor. It may be recalled that 200 MeV of (recoverable) energy is released for every fission event (E_f). Hence, the reactor power is the product of energy released per fission (E_f) and the number of fission events.

Energy released (J/s) = Energy released per fission (E_f)* Number of fissions events per second (fission/s).

Neutron flux(ϕ), defined as the number of neutrons per unit fission cross section per unit time is a key parameter in influencing the number of fission events.

Number of fissions events per second (fission/s)= Neutron flux (neutron/cm^{2}s) * Total fission cross section (cm^2).

Note: Please recall that the fission cross section (σ_f) is expressed per nucleus. Hence the total fission cross section is the fission cross section per nucleus multiplied by the number of fuel nuclei.

Number of fissions events per second (fission/s)= Neutron flux (neutron/cm2s)* Fission cross section per nucleus (cm2)*Number of fuel nuclei.

Therefore,

Energy released (J/s) = Energy released per fission (J)* Neutron flux (neutron/cm^{2}s) * Fission cross section per nucleus (cm^2)*Number of fuel nuclei.

Specific power of the fuel, defined as the thermal energy released per unit mass of the fuel, can be related to the energy released as follows:

Specific power of fuel (W/kg) = Energy released (J/s)/Mass of the fuel (kg).

Substituting the above Equations, we get:

Specific power of fuel (W/kg) = Energy released per fission (J)* Neutron flux (neutron/cm^{2}s) * Macroscopic fission cross section (cm^2)*Number of fuel nuclei/Mass of the fuel (kg).

Let P' be the specific power of fuel, Ef be the energy released per fission in Joule, N_f be the number of fuel nuclei per unit mass of the fuel then,

$$P' = E_f \phi N_f \sigma_f$$

Equation above is a simplified one assuming neutron flux to be independent of neutron energy and radial position. Taking in to account the influence of radial position and neutron energy on the neutron flux, we have:

$$P' = E_f N_f \phi(r,E) \sigma_f$$

Neutron flux corresponding to energy of thermal neutron is the thermal neutron flux, ϕ (r). Hence the specific power of fuel as a function of radial position in the fuel is given by,

$$P' = E_f N_f \phi(r) \sigma_f$$

Power density (P") is defined as the thermal energy released per unit volume of the fuel.

$$P'' = E_f N_f \phi(r) \sigma_f \rho_f$$

If the average neutron flux is ϕ^{-}, then the above Eq. can be written as:

$$P' = E_f N_f \phi^{-} \sigma_f$$

$$P'' = E_f N_f \phi^{-} \sigma_f \rho_f$$

Example – 1: Determine the number of U-235 nuclei in 1 kg of natural UO_2 fuel. The molecular weight of Uranium dioxide is 270.

Solution: From the molecular weight of UO_2 (270) and the atomic weight of Uranium (238), one may calculate the mass of uranium in one kg of UO_2 as follows:

Mass of Uranium in 1 kg of UO_2 = 238/270 = 0.88 kg.

Number of moles of Uranium in 1 kg of UO_2= 0.88/238 = 3.697 gmole.

Recalling the definition of one mole, there are 6.023 x 10^{23} (Avagadro number) atoms or nuclei per mole of a substance.

Therefore, one kg of UO_2 contains 2.227 x 10^{24} atoms.

Please recall that the natural uranium contains only 0.7 % (by mass) of U-235. Hence the number of atoms of U-235 in one kg of UO_2 is the product of atomic fraction of U-235 and 2.227 x 10^{24}.

To convert mass % to atomic fraction, the following method is used.

Atomic % of U-235 = (mass % of U-235/235)/(mass % of U-235/235+ mass % of U- 238/238).

Atomic % of U-235 = (0.7/235)/(0.7/235+99.3/238) = 0.007039.

Note: For natural uranium dioxide fuel, the mass % and atomic % of U-235 are same. However, at higher levels of enrichment the mass % and atomic % may slightly differ.

Therefore, one kg of natural UO_2 contains ~ 2.227 x 10^{24} x 0.007= 1.559 x 10^{22} atoms of U-235.

Example - 2: Determine the specific power and power density of a natural UO_2 fuel in a heavy water reactor. The average neutron flux is 5x10^{13} $cm^{-2}s^{-1}$. The fission cross section is 579 b. The density of UO_2 is 18900 kg/m^3.

Solution:

Writing earlier Eq. again, we have:

$$P' = E_f N_f \phi^{-} \sigma_f$$

E_f = 200 MeV = 3.2 x 10^{-11} J; ϕ^{-}=5x10^{13} $cm^{-2}s^{-1}$; σ_f= 579 b = 579*10^{-24} cm^2

As seen in the example -1, the number of fissile nuclei per kg of natural uranium dioxide is 1.559x10^{22}.

Therefore, P' = 14442 W/kg = 14.44 kW/kg

$$P'' = P' \rho_f$$

P"= 273 MW/m^3

Modes of Heat Transfer

There are three modes by which heat may be transferred from one object to the other:

(i) conduction (ii) convection and (iii) radiation.

Conduction

Conduction is a mode of heat transfer that occurs in solids and in stationary fluid. Conduction occurs due to molecular motion in a substance. Molecules in high temperature region possess high thermal energy by virtue of which they move faster. These molecules collide with adjacent molecules causing them to move as well. This causes energy flow towards the region of lower temperature.

Fourier's law of heat conduction can be used to understand the variables influencing heat transfer by conduction. By Fourier's law, rate of heat transfer (Q) per unit area.

(A) is proportional to the temperature gradient (dT/dx). The constant of proportionality is thermal conductivity of the material.

$$\frac{Q}{A} = -k\frac{dT}{dx} \tag{1}$$

'A' refers to the heat transfer area, measured as the area perpendicular to the direction of heat flow. From above Eq, one can understand that to transfer a certain quantity of heat per unit area, a larger temperature gradient is required for a material with lower thermal conductivity. Hence to promote heat transfer by conduction, use of solid materials with higher thermal conductivity is preferred. However, the materials used to minimize heat losses (insulators) must possess lower thermal conductivity.

Example – 1: A plane wall made of a material with thermal conductivity of 1 W/mK is used as a thermal barrier between a hot chamber and atmosphere. The temperature of the wall on the hotter side is 250 °C while that on the colder side is 30 °C. Determine the rate of heat transfer across the wall, if the wall is 1 m x 1 m x 0.1 m.

Solution: The rate of heat transfer can be calculated using Fourier's law of heat conduction as follows:

$$\frac{Q}{A} = -k\frac{dT}{dx}$$

$$Q = -kA\frac{dT}{dx} \tag{2}$$

k = 1 W/mK; dx = 0.1 m; dT = 220; A = 1 m^2 (Please note that the distance between the hot and cold side is equal to the thickness of the wall).

Substituting the above, in Eq. (2), gives Q = 2200 W.

The rate of heat transfer across the wall is 2200 W.

Thermal conductivity is a characteristic property of a material and is also a function of temperature. The thermal conductivity of a material may be related to temperature as:

$$k = k_0(1 + \beta T) \tag{3}$$

For a material with negative value of 'β' the thermal conductivity of the material.

decreases with temperature, while for materials with positive value of 'β', the thermal conductivity increases with temperature.

When the thermal conductivity of a material is a function of temperature, the rate of heat transfer can be calculated as follows:

$$\frac{Q}{A} = -\frac{1}{L}\int_{T1}^{T2} k(T)dT \qquad ; T_1>T_2 \qquad (4)$$

In Eq. (4), 'L' refers to the thickness ($dx = x_2 - x_1$).

Convection

Convection is a mode of heat transfer associated with the bulk movement of fluid. When a particle of cold fluid comes into contact with a hot surface, it gains heat from the surface. This fluid particles returns to the bulk and mixes with other particles, during which it transfers heat to the bulk and returns to the bulk temperature. With increased contacts between particles of fluid and the hot surface, more heat is transferred from the hot surface to the fluid. Hence, heat transfer by convection depends on the velocity of fluid and the difference in temperature between the hot surface and the cold fluid. If the movement of fluid is achieved by use of external devices like pumps, stirrers or blowers, it is called forced convection. When fluid motion is set up by density difference existing between the hotter and colder parts of the fluid, it is called natural convection. The rate of heat transfer (Q) per unit heat transfer area (A) is related to the difference in temperature between a surface and liquid as,

$$\frac{Q}{A} = h(T_s - T_f)$$

In above Eq, 'h' refers to heat transfer coefficient. Unlike thermal conductivity, heat transfer coefficient is not an inherent property of a material. It is dependent upon the thermo-physical properties of fluid, velocity of the fluid and the geometry of heat transfer system.

Example – 2: A hot cylinder at a temperature of 200 °C is exposed to air at 30°C. If the heat transfer coefficient is 20 W/m²K, determine the rate of heat transfer per unit area.

Solution:

$$\frac{Q}{A} = h(T_s - T_f)$$

h=20 W/m²K; T_s = 200 °C; T_f = 30 °C.

Therefore, the rate of heat transfer per unit is 3400 W/m².

Radiation

Radiation is the mode of heat transfer that involves emission and absorption of electromagnetic radiation between two objects placed at different temperatures. This mode of heat transfer does not require a material medium and can occur even in vacuum. This mode of heat transfer is of less importance in nuclear reactors.

Boiling

Boiling is a phase change (liquid to vapour) heat transfer that occurs when a liquid comes into contact with a solid surface or wall at a temperature (T_w) greater than the saturation temperature of the liquid (T_s). Boiling is characterized by high transfer rates achievable due to the high heat transfer coefficient. Hence boiling is used to remove heat from high-heat flux sources, as in nuclear reactors.

Boiling heat transfer may be classified as pool boiling and convective boiling, depending on whether the fluid is at rest or in motion through channels due to pumps. Convective boiling is also called flow boiling.

Let us discuss flow boiling of a liquid flowing inside a vertical heated tube under constant heat flux conditions, as observed for a boiling liquid in the core of a nuclear reactor. As the liquid well below its boiling point enters the heated tube, heat is carried away from the surface by forced convection. The temperature of liquid increases during this phase. This is similar to single phase heat transfer from a heated surface to a flowing liquid. This regime is between A and B in Figure.

When the temperature of liquid bulk is below the saturation temperature, but the surface is at a temperature above the saturation temperature, vapors are produced closer to the surface and are condensed in the liquid bulk. This is called subcooled boiling (boiling when the temperature of liquid bulk is below saturation temperature). This regime is between B and C in Figure. When subcooled boiling takes place in a vertical channel, a two-phase mixture of bubbles and liquid is present in the channel. This is called bubbly flow.

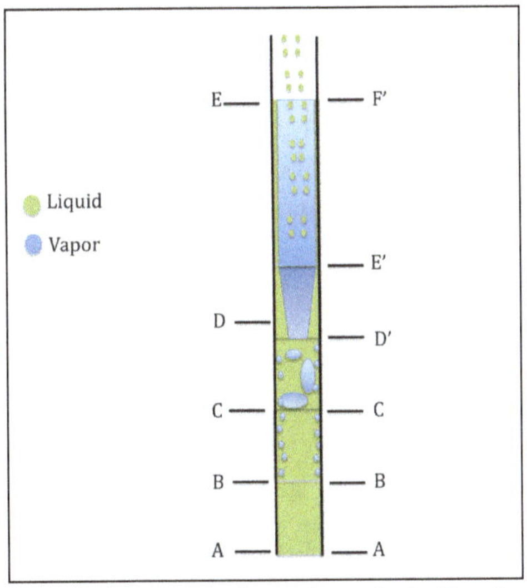

Hydrodynamic and heat transfer regimes for boiling heat transfer.

With the fluid moving further upwards in the channel, the temperature of the liquid increases and reaches the saturation temperature. Under these conditions, the vapors formed on the heated surface and that reach the liquid bulk do not condense. Hence the vapors grow as they reach the liquid bulk. The rate of heat removal from the heated surface is very high and hence the temperature of the surface either remains constant or decreases slowly with height. This regime is called saturated nucleate boiling regime (between C and D). The hydrodynamics of this regime is called slug flow (between C and D').

In pressurized water reactors, the vapor fraction is low and hence the surface is cooled by liquid by nucleate boiling. When the vapors begin to blanket the surface, the rate of heat removal is reduced and a condition called "Departure from Nucleate Boiling (DNB)" results. The "Departure from Nucleate Boiling" has to be minimized to ensure that the temperature of clad and fuel do not increase rapidly.

In boiling water reactors, where the fraction of vapors is high, the surface is cooled by the liquid film (similar to that in annular flow). When liquid is completely out of contact with surface, 'Dry out' results. This is also unfavorable as surface temperature increases rapidly.

Neutron Flux

The neutron flux is a quantity used in nuclear reactor physics corresponding to the total length travelled by all neutrons per unit time and volume, or nearly equivalently number of neutrons travelling through a unit area in unit time. The neutron fluence is defined as the neutron flux integrated over a certain time period.

Natural Neutron Flux

Neutron flux in asymptotic giant branch stars and in supernovae is responsible for most of the natural nucleosynthesis producing elements heavier than iron. In stars there is a relatively low neutron flux on the order of 10^5 to 10^{11} neutrons per cm^2 per second, resulting in nucleosynthesis by the s-process (slow-neutron-capture-process). By contrast, after a core-collapse supernova, there is an extremely high neutron flux, on the order of 10^{22} neutrons per cm^2 per second, resulting in nucleosynthesis by the r-process (rapid-neutron-capture-process).

Atmospheric neutron flux, apparently from thunderstorms, can reach levels of $3 \cdot 10^2$ to $5 \cdot 10^2$ neutrons per cm^2 per sec. However, recent results obtained with unshielded scintillation neutron detectors actually show a decrease in the neutron flux during thunderstorms.

Artificial Neutron Flux

Artificial neutron flux refers to neutron flux which is man-made, either as byproducts from weapons or nuclear energy production or for specific application such as from a research reactor or by spallation. A flow of neutrons is often used to initiate the fission of unstable large nuclei. The additional neutron(s) may cause the nucleus to become unstable, causing it to decay (split) to form more stable products. This effect is essential in fission reactors and nuclear weapons.

Within a nuclear fission reactor the neutron flux is primarily the form of measurement used to control the reaction inside. The flux shape is the term applied to the density or relative strength of the flux as it moves around the reactor. Typically the strongest neutron flux occurs in the middle of the reactor core, becoming lower toward the edges. The higher the neutron flux the greater the chance of a nuclear reaction occurring as there are more neutrons going through an area.

A reactor vessel of a typical nuclear power plant (PWR) endures in 40 years (32 full reactor years) of operation approximately 3.5×10^{19} n/cm^2 (E>1MeV). Neutron flux causes reactor vessels to suffer from embrittlement and the steel gets activated.

Heat Flow Distribution In Plate Fuel Elements

To determine the heat flow distribution, it is essential to study the heat generation in fuel elements as a function of position.

Recalling Eq. (1),

$$\varphi = A\cos\left(\frac{\pi x}{a}\right) = \varphi_{\max}\cos\left(\frac{\pi x}{a}\right) \tag{5}$$

Recalling the relationship between the reactor power per unit volume and neutron flux, we have:

$$P'' = E_f N_f \phi(x) \sigma_f \rho_f \tag{6}$$

P" is reactor power per unit volume (W/m^3); E_f is the recoverable energy released per fission (J); ϕ (x) is the neutron flux as a function of distance from the centre (m^{-2}s^{-1}); N_f is the number of fissile nuclei per unit mass of fuel (kg^{-1}); σ_f is the fission cross section (m^2) and ρf is the density of fuel (kgm^{-3}).

Substituting Eq. (5) in Eq. (6) we get,

$$P'' = E_f N_f \varphi_{\max}\cos\left(\frac{\pi x}{a}\right)\sigma_f \rho_f \tag{7}$$

Equation (7) may be simplified as:

$$P'' = K'\cos\left(\frac{\pi x}{a}\right) \tag{8}$$

where K' = $\phi_{max} E_f N_f \sigma_f \rho_f$

Equation (8) relates the heat generated as a function of distance from the centre of the slab. Equation (1) is similar to Eq. (8) with K' of Eq. (8) replaced by ϕ_{max} of Eq. (5). Hence profile of reactor power as a function of distance from the centre will be qualitatively similar to neutron flux profiles as shown in Figure.

Temperature Distribution in Plate Fuel Elements

The temperature profile in the slab reactor can be obtained by the solution of steady- state energy balance equation with heat generation.

The state-state energy balance equation in Cartesian coordinates with P" as the heat generation rate per unit volume is given by:

$$\frac{\partial^2 T}{\partial x^2}+\frac{\partial^2 T}{\partial y^2}+\frac{\partial^2 T}{\partial z^2}+\frac{P''}{k}=0 \tag{9}$$

Applying Eq. (9) for an infinite slab reactor as one-dimensional steady-state heat conduction, we have,

$$\frac{\partial^2 T}{\partial x^2}+\frac{P''}{k}=0 \tag{10}$$

Substituting Eq. (8) in Eq. (10),

$$\frac{d^2 T}{dx^2}+\frac{K'\cos\left(\frac{\pi x}{a}\right)}{k}=0 \tag{11}$$

Two boundary conditions are required for the solution of Eq. (11).

At the centre of the slab, neutron flux and hence heat generation are maximum. Hence, one would expect the temperature at the centre of the reactor to be the maximum.

Therefore,

$$x=0; \quad \frac{dT}{dx}=0 \tag{12}$$

The other boundary condition is the temperature at the outer surface of the reactor. Let it be T_S.

Therefore,

$$x=\frac{a}{2}; \quad T=T_S \tag{13}$$

Solution of Eq. (11) subjected to the boundary conditions (Eq. (12) & Eq. (13)) leads to,

$$T=T_S+\frac{K'a^2}{k\pi^2}\left(\cos\frac{\pi x}{a}\right) \tag{14}$$

If the heat generated is assumed to be constant with position, or if the average heat generated is taken into account, then Eq. (10) becomes,

$$\frac{d^2 T}{dx^2}+\frac{P''_{avg}}{k}=0 \tag{15}$$

The solution of Eq. (15) subjected to the boundary conditions (Eq. (12) & Eq. (13)) is,

$$T = T_S + \frac{P''_{avg}}{8k}\left(a^2 - x^2\right) \qquad (16)$$

Example 1: A slab of 100 mm thick made up of a material of thermal conductivity 120 W/mK generates heat at a rate of 1.5 MW/m³. Determine the temperature at the centre of the slab, if the outer surface of the slab is maintained at 180 °C.

Data: k = 120 W/mK; T_S = 180 °C; P_{avg}"=1.5x 10^6 W/m³; a = 0.10 m.

Substituting above and x= 0 in Eq. (16), we get T = 196 °C.

Example 2: A slab of 100 mm thick made up of a material of thermal conductivity 120 W/mK generates heat at a rate of 1.5cos(πx/a) MW/m³, where 'x' is the distance from the centre of the slab and 'a' is thickness of the slab. Determine the temperature at the centre of the slab, if the outer surface of the slab is maintained at 180 °C.

Data: k = 120 W/mK; T_S = 180 °C; P"=1.5x 106cos(π x/a); K'= 1.5 x 10^6 W/m³; a = 0.10 m.

Equation (14) is to be used for this purpose:

$$T = T_S + \frac{K'a^2}{k\pi^2}\left(\cos\frac{\pi x}{a}\right)$$

Solving the above equation for x = 0 gives T= 192.7°C.

Equations (14) and (16) can be used to predict the temperature profiles in a slab reactor only if T_S is known. In many instances T_S is not known.

Let us consider the infinite slab to be cooled by water as in the case of a water-cooled reactor. Under these circumstances, the temperature profile in the slab reactor along with its surface temperature can be predicted by equating the heat generated in the slab to the heat removed by coolant from the slab.

Heat generated in the slab per unit volume = P_{avg}"

Total heat generated in the slab = P_{avg}"a LB (17)

In Eq. (17), L and B are the length and breadth of the slab, such that their product is the area perpendicular to the direction of heat flow.

One may recollect the relationship between heat gained by the coolant, heat transfer coefficient and the driving force as,

Heat gained by the coolant = hAΔT (18)

In Eq. (18), 'h' is the heat transfer coefficient, 'A' the heat transfer area (LB) and ΔT, the difference in temperature between hot surface (T_S) and the coolant (T_c).

Therefore,

$$P_{avg}\text{''}a\,LB = h\,A\,(T_S - T_c) \tag{19}$$

Since 'A=LB', we have

$$P_{avg}\text{''}a = h\,(T_S - T_c) \tag{20}$$

Therefore,

$$T_S = T_c + \frac{P^{"}_{avg}a}{h} \tag{21}$$

Example - 3: A slab of thickness 200 mm generates heat at the rate of 200 kW/m³. If the slab is cooled by water at a temperature of 25 °C, determine the temperature on the outer surface of the slab. The heat transfer coefficient may be taken as 400 W/m²K.

Data: T_c = 25 °C, h = 400 W/m²K; P_{avg}"=2x10⁵ W/m³; a = 0.2 m

Substituting the above in Eq. (21), we get T_S = 125 °C.

Power Profile in Finite Cylindrical Reactor

Please recall the neutron flux profile for a cylindrical reactor,

$$\phi = \phi_0 J_0\left(\frac{2.405r}{R'}\right)\cos\left(\frac{\pi z}{H'}\right) \tag{22}$$

Recalling the relationship between reactor power per unit volume (P") and neutron flux (ϕ), we have,

$$P'' = E_f N_f \phi(z,r) \sigma_f \rho_f \tag{23}$$

Substituting Eq. (22) in Eq. (23) we get,

$$P'' = E_f N_f \sigma_f \rho_f \phi_0 J_0 (2.405\frac{r}{R'})\cos(\pi\frac{z}{H'}) \tag{24}$$

Equation (24) may be simplified as:

$$P'' = K'' J_0 (2.405\frac{r}{R'})\cos(\pi\frac{z}{H'}) \tag{25}$$

where K" = $E_f N_f \sigma_f \rho_f$

Equation (25) relates the heat generated as a function of distance from the centre of the cylinder and as function of distance from the bottom. Equation (22) is similar to Eq. (25) with K" of Eq. (25) replaced by ϕ_0 in Eq. (22). Hence profile of reactor power as a function of distance from the centre will be qualitatively similar to neutron flux profiles as shown in Figure.

Temperature Distribution in Cylindrical Fuel Elements

Analogous to the prediction of temperature profile for an infinite slab reactor, the temperature profile in cylindrical fuel elements can also be obtained by the solution of one-dimensional, steady-state energy balance equation with heat generation in cylindrical coordinates. This is shown as Equation (26):

$$\frac{1}{r}\frac{d}{dr}\left(r\frac{dT}{dr}\right)+\frac{P''}{k}=0 \tag{26}$$

To begin with, let us solve Eq. (26) for the case of uniform volumetric rate of heat generation, P''_{avg}

Equation (26) becomes:

$$\frac{1}{r}\frac{d}{dr}\left(r\frac{dT}{dr}\right)+\frac{P''_{avg}}{k}=0 \tag{27}$$

The temperature will be maximum at the centre of the fuel element. Hence one of the boundary conditions is:

At r = 0; dT/dr=0 (28)

Let the temperature on the outer surface of the fuel be T_S. Accordingly, the second boundary condition is:

At r = R; T = T_S (29)

To solve Eq. (27) using the boundary conditions given in Eq. (28) and Eq. (29), we may rearrange Eq. (27) as follows:

$$\frac{d}{dr}\left(r\frac{dT}{dr}\right)+\frac{rP''_{avg}}{k}=0 \tag{30}$$

Integrating Eq. (30),

$$r\left(\frac{dT}{dr}\right)=\frac{-r^2P''_{avg}}{2k}+C_1 \tag{31}$$

Applying the first boundary condition (r =0; dT/dr = 0) gives C_1 = 0 and Eq. (31) becomes,

$$r\left(\frac{dT}{dr}\right)=\frac{-r^2P''_{avg}}{2k} \tag{32}$$

Integrating Eq. (32), we get:

$$T=\frac{-r^2P''_{avg}}{4k}+C_2 \tag{33}$$

Applying the second boundary condition (r = R; T = T_S), we get:

$$C_2 = T_S + \frac{R^2 P''_{avg}}{4k} \tag{34}$$

Substitution of Eq. (34) in Eq. (27) gives:

$$T = T_S + \frac{R^2 P''_{avg}}{4k} - \frac{r^2 P''_{avg}}{4k} \tag{35}$$

Re-arranging Eq. (35),

$$T = T_S + \frac{R^2 P''_{avg}}{4k}\left(1 - \frac{r^2}{R^2}\right) \tag{36}$$

The surface temperature (at the outer surface of the cylinder) can be obtained by equating the heat generated in the cylinder with the heat removed by the coolant as follows (similar to how it was done for slab geometry),

$$P''_{avg} H \frac{\pi D^2}{4} = h\pi DH(T_S - T_c) \tag{37}$$

Note: π DH is the heat transfer area or the area perpendicular to the direction of heat transfer. π $D^2/4$ is the cross sectional area which when multiplied by the height of the cylinder (H) gives its volume. T_c is the temperature of the coolant.

Rearrangement of Eq. (37) gives,

$$T_S = T_c + P''_{avg} \frac{D}{4h} \tag{38}$$

The maximum temperature (centre-line temperature) is obtained by substituting r=0 and Eq. (38) in Eq. (36) as follows:

$$T = T_c + P''_{avg} \frac{D}{4h} + \frac{R^2 P''_{avg}}{4k} \tag{39}$$

Example – 1: A cylindrical fuel rod of diameter 1.1 cm generates heat at the rate of 10 MW/m³. If water at a temperature of 220 °C extracts the heat from the fuel rod, determine the maximum temperature of the fuel rod and the surface temperature of the fuel rod. The heat transfer coefficient may be taken as 1000 W/m²K. The thermal conductivity of fuel is 30 W/mK.

Data: T_c = 220°C; P_{avg}"=10^7 W/m³; D = 0.011 m; h = 1000 W/m²K; k = 30 W/mK.

Before estimating the maximum temperature, the temperature on the outer surface of the rod (T_S) can be calculated using Eq. (38):

$$T_S = 247.5\ °C$$

Now, Eq. (15) can be used with r = 0 and T_S=247.5 to determine the maximum temperature as:

$$T_{max} = 250\ °C.$$

(Note: The lower temperature difference between the temperature at the centre of the rod and that at the outer surface is low due to smaller diameter of the rod).

Temperature Distribution in Cylindrical Fuel Elements with Position Dependent flux

Recall Eq. (24):

$$P'' = E_f N_f \sigma_f \rho_f \phi_0 J_0 (2.405\frac{r}{R'})\cos(\pi\frac{z}{H'})$$

At a fixed axial position, P" is dependent on 'r' alone may be expressed as:

$$P'' = K''' r/R = P_{max}'' r/R \tag{40}$$

Substituting Eq. (40) in Eq. (27), we have:

$$\frac{1}{r}\frac{d}{dr}\left(r\frac{dT}{dr}\right)+\frac{P''_{max} r}{kR}=0 \tag{41}$$

Eq. (41) can be solved as follows:

$$\frac{d}{dr}\left(r\frac{dT}{dr}\right)+\frac{P''_{max} r^2}{kR}=0 \tag{42}$$

Integrating Eq. (42), we get:

$$\left(r\frac{dT}{dr}\right)=-\frac{P''_{max} r^3}{3kR}+C \tag{43}$$

Applying the first boundary condition (r =0; dT/dr=0),

$$C=0 \tag{44}$$

Therefore, Eq. (43) becomes:

$$\left(\frac{dT}{dr}\right)=-\frac{P''_{max} r^2}{3kR} \tag{45}$$

Integrating Eq. (45),

$$T = -\frac{P''_{max} r^3}{9kR} + C' \tag{46}$$

Applying the second boundary condition (r= R; T = T_s),

$$C' = T_s + \frac{P''_{max} R^3}{9kR} \tag{47}$$

Substituting Eq. (47) in Eq. (46),

$$T = T_s + \frac{P''_{max} R^3}{9kR} - \frac{P''_{max} r^3}{9kR} \tag{48}$$

$$T = T_s + \frac{P''_{max} R^3}{9kR}\left(1 - \frac{r^3}{R^3}\right) \tag{49}$$

Equation (38) holds good for the determination of 'T_s'.

Temperature Distribution in Cylindrical Fuel Elements with Temperature Dependent Thermal Conductivity and Uniform Heat Generation

So far, we have assumed thermal conductivity to be independent of temperature. However, thermal conductivity of solids is a function of temperature.

Recalling Eq. (24),

$$k = k_0(1 + \beta T) \tag{50}$$

Recalling Eq. (26),

$$\frac{1}{r}\frac{d}{dr}\left(r\frac{dT}{dr}\right) + \frac{P''}{k} = 0 \tag{51}$$

Substituting Eq. (50) in Eq. (51) results in,

$$\frac{1}{r}\frac{d}{dr}\left(r\frac{dT}{dr}\right) + \frac{P''}{k_0(1+\beta T)} = 0 \tag{52}$$

Rearranging Eq. (52),

$$\frac{d}{dr}\left(r\frac{dT}{dr}\right) + \frac{rP''}{k_0(1+\beta T)} = 0 \tag{53}$$

Integrating Eq. (53),

$$\left(r\frac{dT}{dr}\right) = -\frac{r^2 P''}{2k_0(1+\beta T)} + C \tag{54}$$

At r = 0; dT/dr = 0

Therefore, C = 0 and Eq. (54) becomes,

$$\left(\frac{dT}{dr}\right) = -\frac{rP''}{2k_0(1+\beta T)} \tag{55}$$

Integrating Eq. (55),

$$\int(1+\beta T)dT = -\int\frac{rP''}{2k_0}dr \tag{56}$$

$$T + \frac{\beta T^2}{2} = -\frac{r^2 P''}{4k_0} + C' \tag{57}$$

Substituting the second boundary condition, (r = R; T = T_s),

$$C' = T_s + \frac{\beta T_s^2}{2} + \frac{R^2 P''}{4k_0} \tag{58}$$

Substitution of Eq. (58) in Eq. (57) gives,

$$T + \frac{\beta T^2}{2} = T_s + \frac{\beta T_s^2}{2} + \frac{R^2 P''}{4k_0}\left(1 - \frac{r^2}{R^2}\right) \tag{59}$$

Equation (59) is the expression for temperature as a function of radial position. If the thermal conductivity is taken to be independent of temperature, β =0. Substituting β =0 in Eq. (59) gives,

$$T = T_s + \frac{R^2 P''}{4k_0}\left(1 - \frac{r^2}{R^2}\right) \tag{60}$$

Please note that Eq. (60) is same as that of Eq. (57).

Equation (59) is a quadratic equation, solution of which is given below:

$$T = \frac{-1 + \sqrt{1 + 2\beta\left\{T_s + \beta\frac{T_s^2}{2} + \frac{P''R^2}{4k_0}\left(1 - \frac{r^2}{R^2}\right)\right\}}}{\beta} \tag{61}$$

In which temperature distribution within a cylindrical fuel element was studied. The analysis of temperature distribution will include cladding also.

The learners will be able to estimate the temperature distribution in cylindrical fuel element and cladding through steady-state energy balance.

It may be recalled that the cladding is used to prevent direct contact between the coolant and fuel. Hence cladding presents an additional resistance for heat transfer between coolant and fuel. Also a small gap exists between fuel and cladding called fuel-clad gap for accommodation of fission gases.

Let us analyze the temperature distribution in cylindrical fuel and the cladding using steady state energy balance in cylindrical coordinates. Figure shows the geometry considered for the analysis.

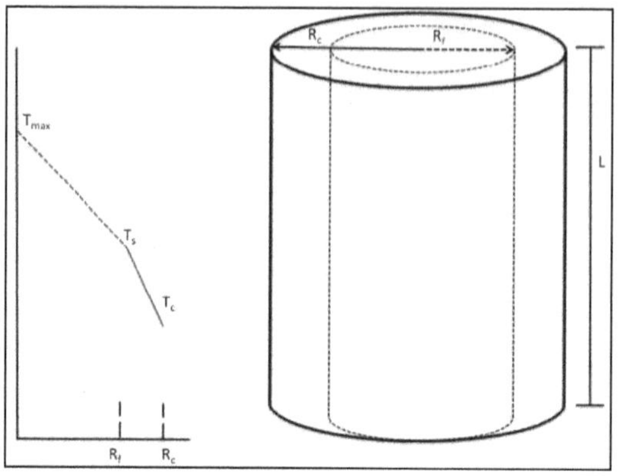

Geometry of cylindrical fuel element with cladding for analysis.

Nomenclature

R_f, R_c – Radii of fuel and cladding respectively.

T_f, T_c – Temperature on the outer surface of fuel and cladding respectively.

T_{max} – Maximum temperature (at the centre of the fuel).

T_∞ - Temperature of the coolant.

k_f, k_c – Thermal conductivities of fuel and cladding material respectively.

Determination of T_c

In Figure, there are three unknown temperatures: T_{max}, T_f and T_c. It is easier to calculate the T_c, the temperature at the outer surface of the cladding by making a simple energy balance.

Under steady-state conditions, the rate of heat transfer is same across the fuel layer and cladding layer. This is also same as the rate of heat transfer from cladding to the coolant.

Heat transfer rate from cladding to the coolant = $h(T_c-T_\infty)\,\pi\, R_c L$ (62)

If P" is the volumetric rate of heat generation in the fuel, then the rate of heat generation for the entire fuel volume is 'Q':

$$Q = P''\pi R_f^2 L \quad (63)$$

This rate of heat generation in the fuel has to be ultimately removed by the coolant. Hence Equating (62) & (63),

$$Q = P''\pi R_f^2 L = h(T_c-T_\infty)\pi R_c L \quad (64)$$

Therefore,

$$T_c = T_\infty + \frac{P"R_f^2}{hR_c} \tag{65}$$

Equation (65) can be used to determine the temperature on the outer surface of the cladding.

Determination of T_s

The rate of heat transfer across any surface is the ratio of driving force to heat transfer resistance (R_T).

Accordingly, heat transfer across cladding layer is equal to ratio of temperature difference across the cladding layer to the heat transfer resistance in cladding layer.

Recalling Fourier's law of heat conduction,

$$\frac{Q}{A} = -k\frac{dT}{dx} \tag{66}$$

Rewriting Eq. (66) for an elemental cylindrical volume of Length (L) and radius (r), we have:

$$Q = -k(2\pi rL)\frac{dT}{dr} \tag{67}$$

Eq. (67) can be integrated with the following limits:

At r = R_f, T = T_f;

&

At r = R_c, T= T_c;

The solution is shown below:

$$Q = \frac{2\pi Lk_c(T_f - T_c)}{ln\left(\frac{R_c}{R_f}\right)} \tag{68}$$

Rearranging Eq. (68), we have:

$$T_f = T_c + \frac{Qln\left(\frac{R_c}{R_f}\right)}{2\pi Lk_c} \tag{69}$$

Temperature (T) at any location (r) in the fuel can be determined as follows, from the knowledge of temperature of the fuel (T_f) on its outer surface (R_f):

$$T = T_f + \frac{R_f^2 P"}{4k_f}\left(1 - \frac{r^2}{R_f^2}\right) \tag{70}$$

Substituting Eq. (69) in Eq. (70),

$$T = T_c + \frac{Q\ln(R_c/R_f)}{2\pi L k_c} + \frac{R_f^2 P''}{4k_f}\left(1 - \frac{r^2}{R_f^2}\right) \quad (71)$$

Equation (71) can be used to predict temperature profile in the cylindrical fuel element.

Substituting r=0 in Eq. (71), one can determine the maximum fuel temperature as given below:

$$T_{max} = T_c + \frac{Q\ln(R_c/R_f)}{2\pi L k_c} + \frac{R_f^2 P''}{4k_f} \quad (72)$$

Example – 1: A cylindrical fuel rod of 8 mm radius made of a material with thermal conductivity 0.7 W/mK generates heat at the rate of 1000 kW/m³. The fuel rod is surrounded by a cladding layer of thickness 4 mm. The thermal conductivity of cladding layer is 1.5 W/mK. If the cladding is cooled by coolant at 30 °C, determine the maximum temperature in the fuel, temperature at the outer surface of the fuel and temperature at the outer surface of the cladding. Heat transfer coefficient may be taken as 900 W/m²K. The length of fuel rod is 1 m.

Data:

$R_f = 8\text{x}10^{-3}$ m; $R_c = R_f$+clad thickness = 12×10^{-3} m.

$k_f = 0.7$ W/mK; $k_c = 1.5$ W/mK.

h = 900 W/m²K.

$T_\infty = 30$ °C.

P" = 1×10^6 W/m³.

Step – I: Determine T_c using Eq. (4):

$$T_c = T_\infty + \frac{P''R_f^2}{hR_c}$$

$T_c = 30 + 1e6*(8\times10^{-3})^2/(900*12\times10^{-3}) = 35.93$ °C.

Step – II: Determine Q:

$Q = P''\pi R_f^2 L = 1\times10^6*3.14*(8\times10^{-3})^2*1 = 200.96$ W.

Step – III: Determine T_f using Eq. (69):

$$T_f = T_c + \frac{Q\ln\left(\frac{R_c}{R_f}\right)}{2\pi L k_c}$$

$$T_f = 35.93 + \frac{200.96\ln(12/8)}{2\pi*1.5} = 44.58$$

Step – IV: Determine maximum fuel temperature (T_{max}) from Eq. (72):

$$T_{max} = T_c + \frac{Q\ln(R_c/R_f)}{2\pi L k_c} + \frac{R_f^2 P''}{4k_f}$$

T_{max} = 67.44 °C.

Temperature Distribution in Cylindrical Fuel Elements with Position Dependent flux

The volumetric rate of heat generation per unit volume of fuel is related to radial position in the fuel as follows:

$$P'' = K'''r/R_f = P_{max}''r/R_f \tag{73}$$

T_c has to be determined to begin with. However, calculation of T_c requires heat transfer rate, which in turn requires average volumetric rate of heat generation.

The average volumetric rate of heat generation P_{avg}'' can be determined as follows:

$$P''_{avg} = \frac{1}{\pi R_f^2}\int_0^{R_f} P''\, 2\pi r dr = \frac{1}{\pi R_f^2}\int_0^{R_f} P''_{max}\,\frac{r}{R_f}\,2\pi r dr \tag{74}$$

$$P''_{avg} = \frac{2P''_{max}}{R_f^3}\int_0^{R_f} r^2 dr = \frac{2P''_{max}}{R_f^3}\left(\frac{R_f^3}{3}\right) \tag{75}$$

Therefore,

$$P_{avg}'' = 2P_{max}''/3 \tag{76}$$

Replacing P” in Eq. (65) with P_{avg}'', we get:

$$T_c = T_\infty + \frac{P''_{avg} R_f^2}{hR_c} \tag{77}$$

Rate of heat transfer becomes:

$$Q = P_{avg}''\,\pi\, R_f^2 L \tag{78}$$

The temperature on the surface of the fuel rod can be determined using Eq. (69). The maximum temperature at the centre of the fuel can be determined using Eq. (72) by replacing (P”) of Eq. (72) and P_{avg}'' as follows:

$$T_{max} = T_c + \frac{Q\ln(R_c/R_f)}{2\pi L k_c} + \frac{R_f^2 P''_{avg}}{4k_f} \tag{79}$$

Example – 2: A cylindrical fuel rod of 20 mm radius made of a material with thermal conductivity 1 W/mK generates heat at the rate of $3000(r/R_f)$ kW/m³. The fuel rod is surrounded by a cladding layer of thickness 5 mm. The thermal conductivity of cladding layer is 3 W/mK. If the cladding is cooled by coolant at 25 °C, determine the maximum temperature in the fuel, temperature at the outer surface of the fuel and temperature at the outer surface of the cladding. Heat transfer coefficient may be taken as 1000 W/m²K. The length of fuel rod is 1 m.

Step – I: Determine average volumetric rate of heat generation using Eq. (76):

$P_{avg}'' = 2P_{max}''/3 = 2000 \text{ kW/m}^3$.

Using Eq. (16), T_c can be calculated as follows:

$$T_c = T_\infty + \frac{P''_{avg} R_f^2}{hR_c}$$

$T_c = 57$ °C.

Rate of heat transfer (Q) can be found out using Eq. (17):

$Q = P_{avg}'' \pi R_f^2 L$ = 2000e3*3.14*(20e-3)^2*1 = 2512 W.

Using Eq. (8), T_f can be calculated as shown below:

$$T_f = T_c + \frac{Q\ln\left(\frac{R_c}{R_f}\right)}{2\pi L k_c}$$

$T_f = 86.75$ °C.

Using Eq. (79), the maximum temperature at the centre of the fuel rod can be calculated as:

$$T_{max} = T_c + \frac{Q\ln(R_c/R_f)}{2\pi L k_c} + \frac{R_f^2 P''_{avg}}{4k_f};$$

$T_{max} = 286.75$ °C.

Steam Generator

The inverted U-tube bundle of a Combustion Engineering steam generator.

Steam generators are heat exchangers used to convert water into steam from heat produced in a nuclear reactor core. They are used in pressurized water reactors (PWR) between the primary and secondary coolant loops.

In other types of reactors, such as the pressurised heavy water reactors of the CANDU design, the primary fluid is heavy water. Liquid metal cooled reactors such as the Russian BN-600 reactor also use heat exchangers between primary metal coolant and at the secondary water coolant.

Boiling water reactors (BWR) do not use steam generators, as turbine steam is produced directly in the reactor core. Activation of oxygen and dissolved nitrogen in the water means that the turbine hall is inaccessible during reactor operation and for some time afterwards.

Description

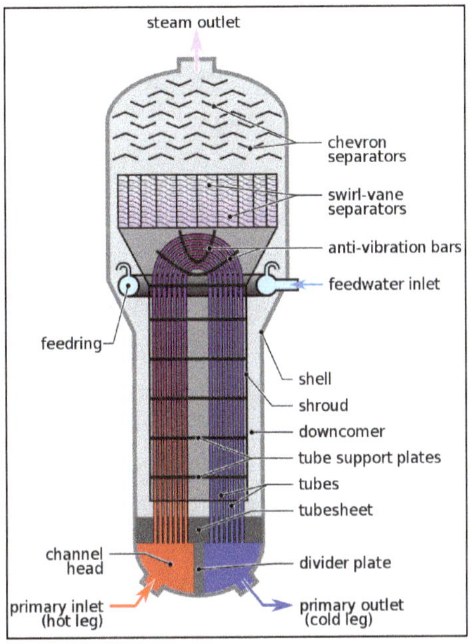

Vertical recirculating-type steam generator (typical of Westinghouse and Combustion Engineering-designed reactors) and components.

In commercial power plants, there are two to four steam generators per reactor; each steam generator can measure up to 70 feet (21 m) in height and weigh as much as 800 tons. Each steam generator can contain anywhere from 3,000 to 16,000 tubes, each about .75 inches (19 mm) in diameter. The coolant (treated water), which is maintained at high pressure to prevent boiling, is pumped through the nuclear reactor core. Heat transfer takes place between the reactor core and the circulating water and the coolant is then pumped through the primary tube side of the steam generator by coolant pumps before returning to the reactor core. This is referred to as the primary loop.

That water flowing through the steam generator boils water on the shell side (which is kept at a lower pressure than the primary side) to produce steam. This is referred to as the secondary loop. The secondary-side steam is delivered to the turbines to make electricity. The steam is subsequently condensed via cooled water from a tertiary loop and returned to the steam generator to be heated once again. The tertiary cooling water may be recirculated to cooling towers where it sheds waste heat before returning to condense more steam. Once-through tertiary cooling may otherwise be provided by a river, lake, or ocean. This primary, secondary, tertiary cooling scheme is the basis of the pressurized water reactor, which is the most common way to extract usable energy from a controlled nuclear reaction.

These loops also have an important safety role because they constitute one of the primary barriers between the radioactive and non-radioactive sides of the plant as the primary coolant becomes radioactive from its exposure to the core. For this reason, the integrity of the tubing is essential in minimizing the leakage of water between the primary and secondary sides of the plant. Steam generator tubes often degrade over time. There is the potential that, if a tube bursts while a plant is operating, contaminated steam could escape directly to the secondary cooling loop. Thus during scheduled maintenance outages or shutdowns, some or all of the steam generator tubes are inspected by eddy-current testing, and individual tubes can be plugged to remove them from operation.

Entire steam generators are often replaced in plant mid-life, which is a major undertaking. Most U.S. PWR plants have had steam generators replaced.

History

The nuclear powered steam generator started as a power plant for the first nuclear submarine, the USS *Nautilus* (SSN-571). It was designed and built by the Westinghouse power company for the submarine from there the company started its development and research of nuclear powered steam generators. Once peaceful nuclear reactors were legalized for use as power plants, power corporations jumped at the opportunity to utilize the growing development of nuclear powered steam generators. Westinghouse built one of the first nuclear power plants, the Yankee Rowe nuclear power station (NPS), which also used a nuclear powered steam generator, in 1960. This power plant had a one hundred MWe (mega watt electric) output. By comparison, some modern plants have over 1100 MWe output. Eventually, other international companies such as Babcock & Wilcox and Combustion Engineering began their own programs for research and development of the nuclear power steam generator. Since the 1960s, the US has fallen behind on some European nations in embracing this new power source. France and the UK have been more actively pursuing the benefits that come with nuclear energy, while the US is more concerned about the risk. Finally, it seems China is planning a massive increase to their nuclear power supply and ordering many new plants to be built.

Three Mile Island

In the Three Mile Island disaster, a main feed water pump shut down, although the cause is not known. Without that pump, the steam generator wasn't able to remove heat from the reactor, so pressure in the reactor began to rise. The system automatically began to dump water from the reactor to reduce pressure, but the relief valve got stuck open when the automation told it to close. The control room indicated that the valve was closed. The staff, therefore, had no idea that they were dumping radioactive water out of one of their reactors. With the water being pumped out, there wasn't enough emergency cooling water and the staff was unaware that they were losing more by the minute. Without adequate cooling, one of the reactors began to melt. The pipes burst and about half the core melted during the accident. Unlike Chernobyl and other nuclear disasters, the containment house around the reactor held and the damage to the outside world was very minimal. Since the containment housing held, no radioactive particles were released into the atmosphere, like what happened in Fukushima. The release of radioactive water did damage and contaminate the local area, but it did not spread from there.

Types

The giant vessel moves in a special train with speed restricted to 20 miles per hour.

This Babcock & Wilcox nuclear steam generator moved in a special train (restricted to 20 mph) via the Penn Central Railroad and Southern Railway from Barberton, Ohio to a Duke Energy site in Oconee, S.C. This generator weights 1,140,000 lbs and is a record shipment for the PC at that time (1970).

Westinghouse and Combustion Engineering designs have vertical U-tubes with inverted tubes for the primary water. Canadian, Japanese, French, and German PWR suppliers use the vertical configuration as well. Russian VVER reactor designs use horizontal steam generators, which have the tubes mounted horizontally. Babcock & Wilcox plants (e.g., Three Mile Island) have smaller steam generators that force water through the top of the OTSGs (once-through steam generators; counter-flow to the feedwater) and out the bottom to be recirculated by the reactor coolant pumps. The horizontal design has proven to be less susceptible to degradation than the vertical U-tube design.

Materials and Construction

The materials that make up the turbine and pipes of a nuclear powered steam generator are specially made and specifically designed to withstand the heat and radiation of the reactor. The water tubes also have to be able to resist corrosion from water for an extended period of time. The pipes that are used in American reactors are made of Inconel, either Alloy 600 or Alloy 690. Alloy 690 is made with extra chromium and most facilities heat treat the metal to make it better able to resist heat and corrosion. The high nickel content in Alloy 600 and Alloy 690 make them well suited for resisting acids and high degrees of stress and temperature.

Degradation

The annealed, or heat treated, Alloy 600 was prone to tube denting and thinning due to water chemistry. Plants that used the Alloy 600 in their water tubes therefore had to install new water chemistry controllers and change the chemicals they put in the water. Due to this, pipe thinning has been taken care of, but on rare occasions, tube denting still occurs, causing leaks and ruptures. The only way to prevent this is regular maintenance and check-ups, but this forces the reactor to shut down. In some cases, plants replaced their Alloy 600 tubes with Alloy 690 tubes and a few plants were shut down. To prevent future problems, manufacturers of steam turbines for nuclear power plants have improved their fabrication techniques and used other materials, such as stainless steel, to prevent tube denting.

Safety Devices

As the disaster at Three Mile Island could have been avoided with prior planning and a water level indicator, the Nuclear Regulatory Commission has started to push for water level controllers. The controller would regulate water to the reactor using a combination of sensors such as feedback controllers and feed-forward controllers. Yet this is only one of many such devices that ensure the safe and efficient production of power by nuclear reaction. Other tools include the control rods, relief valves, and even back up cooling systems. The control rods work by reducing the amount of radiation produced. They are built of a material that absorbs neutrons, and therefore reduces the number of fission reactions that take place inside the reactor. Relief valves function to vent pressure, sometimes into the atmosphere, in order to protect the system as a whole. And if the worst were to happen and a meltdown was occurring, a reservoir of cooling water waiting in standby could mean the difference between a disaster and a minor incident.

Typical Operating Conditions

Steam generators in a "typical" PWR in the USA have the following operating conditions:

Side	Pressure (absolute)	Inlet temperature	Outlet temperature
Primary side (tube side)	15.5 MPa (2,250 psi)	315 °C (599 °F) (liquid water)	275 °C (527 °F) (liquid water)
Secondary side (shell side)	6.2 MPa (900 psi)	220 °C (428 °F) (liquid water)	275 °C (527 °F) (saturated steam)

Tube Material

Various high-performance alloys and superalloys have been used for steam generator tubing, including type 316 stainless steel, Alloy 400, Alloy 600MA (mill annealed), Alloy 600TT (thermally treated), Alloy 690TT, and Alloy 800Mod.

Heat Transport

Shutdown of a nuclear reactor refers to termination of fission chain reaction. A nuclear reactor may be shutdown for one of the following reasons:

(i) The reactor is considered to have served for a time period for which it was designed.

(ii) The reactor requires repairs.

(iii) The reactor has suffered from a severe operational transient and must be prevented from suffering a severe accident.

A reactor is shutdown by inserting the control rods completely into the core or by introducing poisons like boron or both. Despite the termination of fission chain reaction, a considerable amount of heat is still present in the fuel rods in the form of sensible heat. Apart from this, heat is also generated due to decay of fission products and also due to fissions induced by delayed neutrons. Hence the nuclear reactor must be cooled for several hours even after shut down.

Let us look at the quantity of heat released after shutdown as a function of time after shutdown.

Fissions induced by delayed neutrons:

The heat released due to fissions induced by delayed neutrons decreases exponentially with time. In about 80-100 seconds, the heat released as a percentage of reactor power deceases to less than 2 %. Hence this constitutes only a small portion of shutdown cooling and can be removed in a short period of time.

Decay of fission products:

The heat released due to decay of fission products (also called decay heat) is prominent over a longer time period and hence determines the cooling requirements. The decay heat depends upon a number of factors including the number of days in which the reactor was in operation before shutdown, type of reactor etc.

This is approximately 7 % of reactor power (immediately after shutdown), 4 % after 100 seconds, 1.3 % after one hour, 0.5 % after 28 hours and 0.2 % after a week.

The normal heat removal system, if functioning properly, is sufficient for decay heat removal with a small change in the flow path. Under normal reactor operation, steam passes through the turbine and then to condenser before being sent back to core. However during decay heat removal, steam is directly sent to condenser bypassing the turbine. The condensate leaving the condenser is supplied back to the core, in case of BWR and to steam generator in case of PWR.

Shutdown Cooling System in BWR

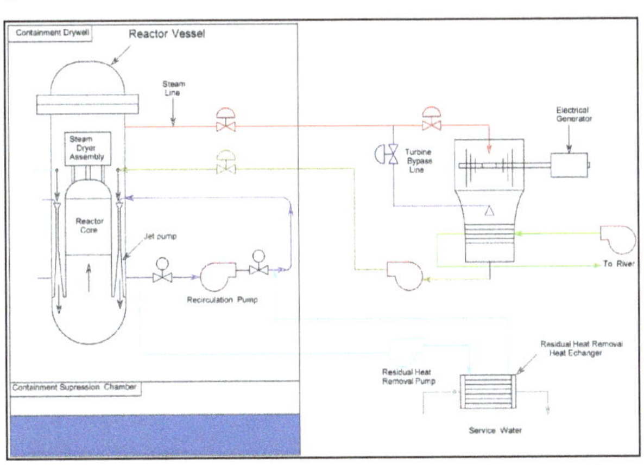

Schematic layout of shutdown cooling system of a typical BWR. The Residual Heat Removal System is also shown (Redrawn from Ref.)

A schematic diagram of shutdown cooling system in a typical Boiling Water Reactor is shown in Figure. In some boiling water reactors (similar to the one at Tarapur), an emergency condenser is provided which comes to operation in cases where the main condenser is isolated from the reactor. The emergency condenser has a huge pool of water that can act as heat sink to remove decay heat in a passive manner (by natural convection) for a period of about 8 hours. BWRs house a Residual heat removal system in which the hot liquid from core is pumped through an external heat exchanger and then returned back to core. This is utilized when the decay heat is insufficient for steam generation.

Shutdown Cooling System in PWR

A schematic diagram of shutdown cooling system in a typical Pressurized Water Reactor is shown in Figure.

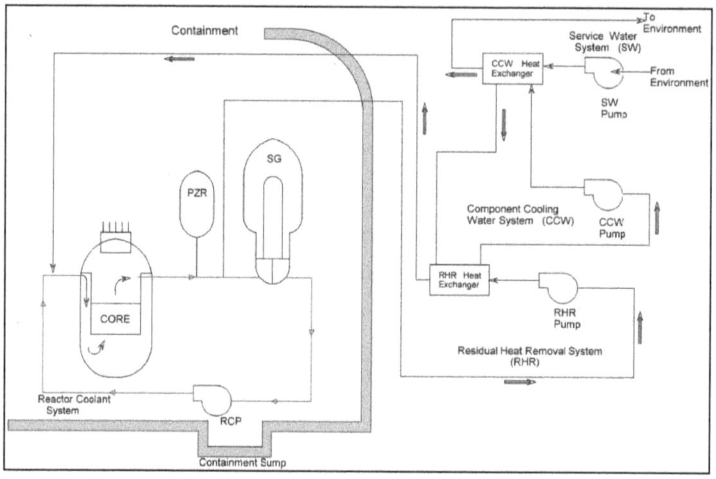

Schematic layout of shutdown cooling system of a typical PWR. The Residual Heat Removal System is also shown (Redrawn from Ref.)

In PWR, during decay heat removal, there are provisions for steam to be vented to atmosphere through atmospheric relief valves, in cases where cooling water for condenser is unavailable. When the decay heat is too low to generate sufficient steam, the Residual Heat Removal (RHR) system is used. In this system, a part of the primary coolant leaving the core is circulated through a heat exchanger to cool the primary coolant to the inlet temperature. This system brings the temperature to a very low value enabling personnel access.

Emergency Core Cooling System

Emergency core cooling system refers to the cooling system adopted in the cases of emergency. This could arise to substantial break in the primary coolant circuit that leads to the decrease in coolant pressure. ECCS becomes operative when the pressure in the primary coolant loop falls below the desired value.

ECCS contains several systems that function independent of each other. Each of these systems is designed to be activated at different pressure levels. All these systems introduce water that is supplied from different sources. Considerable amount of redundancy exists in these systems to ensure the reliability of ECCS. The arrangement of ECCS is different for different reactor types, taking into account of the core thermal-hydraulic characteristics and reactor physics aspects.

Let us discuss ECCS in pressurized water reactors, before we proceed with discussion on ECCS in boiling water reactors.

ECCS in Pressurized Water Reactors

The primary role of ECCS in PWR is of two fold: (i) To provide cooling to the core in order to minimize damage to the fuel, in the case of an accident caused by the loss of coolant. For this purpose, large amounts of low temperature, borated water is injected into the reactor (ii) To ensure that the reactor remains shutdown after the reactor has been cooled following a rupture of steam line. This is done by pumping water containing excess of neutron poisons stored in Refueling Water Storage Tank (RWST).

The Emergency Core Cooling System of a Pressurized water reactor consists of three subsystems. They are (i) High presssure and intermediate pressure injection system; (ii) accumulator injection system; (iii) low-pressure injection system. Each subsystem gets activated depending on the specific signals from the reactor say, coolant pressure. Each subsystem is provided with 2 pumps of equal capacity and capable of delivering the required flow individually. This is done to maintain redundancy. These pumps are operated using diesel generators and hence can operate even if the normal source of electricity for reactor is unavailable.

High Pressure and Intermediate Injection Systems

High-pressure injection system comes into operation in the case of moderate drop in pressure say from 155 bar to 110 bar. There are two conditions that lead to such a situation:

(a) Small break in primary coolant circuit: This will result in reduced coolant flow rate. If un-noticed, this will result in appreciable boiling of coolant in the core which is undesirable in pressurized water reactors.

(b) Failure of pressure relief valve in the pressurizer to close after being activated in an over-pressurization situation: Such a scenario may result in the loss of primary coolant, resulting in the boiling of coolant in core. It is relevant to talk about the "Three-Mile Island" accident at this juncture. In the reactor – 2 of the Three Mile Island plant, the pilot-operated relief valve was opened due to increase in steam pressure caused by reduced primary coolant flow. However, upon reaching the pressure close to normal operational limits, the valve must have closed. However, a mechanical fault led to non-closure of the relief valve. This valve in 'open position' allowed the escape of large quantities of water from the primary system, ultimately leading to fuel melt down.

Upon actuated, the high pressure injection system works by injecting borated water into reactor coolant inlet pipelines called cold legs, through the use of pumps driven by electricity. These pumps are the ones that are used in chemical and volume control system (CVCS). The acutation of this subsystem makes the pumps take input from refueling water storage tank (RWST) instead of from CVCS. Intermediate pressure system is activated in cases of moderate to intermediate size breaks in the primary coolant loop. This system too takes water from RWST. Hence high pressure and intermediate injection systems are sometimes considered together as a single subsystem in some cases.

Accumulator Injection System

In the event of a large break in primary coolant circuit, system pressure decreases rapidly due to which another ECCS subsystem called 'Accumulator injection system' gets actuated. The system contains large tanks called 'Accumulators' that contain cool borated water under nitrogen gas at a pressure of about 14 to 40 bar. These tanks are connected to cold leg (primary coolant inlet) via a non-return valve that gets actuated without manual intervention when the pressure in the cold leg is below that in accumulators. Hence this is a "passive system" that depends on gravity for discharge of borated water into the cold leg without the use of pumps.

Low-pressure Injection System

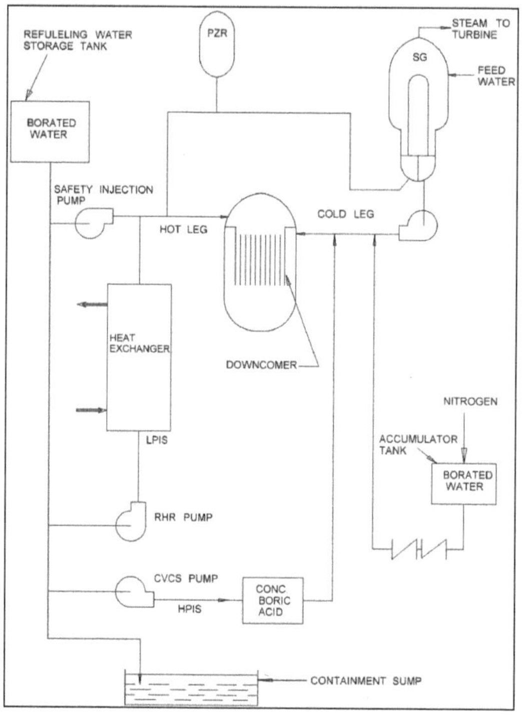

Schematic diagram of ECCS of a typical PWR (Redrawn from Ref.)

The Low-pressure injection system is also called the Residual Heat Removal system. In case of large breaks in primary coolant, the coolant pressure will go down to lower values acutating this subsystem. The pumps in this subsystem are also electrically operated and hence require AC power or power from diesel generators. Water is normally drawn from refuelling storage water tank. In case the water in RWST is exhasuted, required water may also be drawn from a sump located in the containment structure. Such water after passing through the core and extracting the heat, is cooled using a residual heat removal system heat exchanger and returns back to cool the reactor.

ECCS in Boiling Water Reactors

ECCS in BWR consists of three independent sub-systems.

High Pressure Coolant Injection Systems

These inject large amount of water into the core to ensure that the core is covered with water. This system can inject water to the core even at high pressures. Under normal circumstances, core is always covered with water. However, breakage of coolant pipes may lead to loss of coolant, which if allowed to proceed without intervention would result in the exposure of core. In case of water level falling below safe limit, other sub-systems come to fore.

With the advancement in BWR technology, minor variations in this ECCS subsystem have been made. For example, BWR-6 manufactured by GE, uses a High-Pressure Core Spray, located between steam separator and top portion of the core, to spray water on to the core.

Automatic Depressurization System (ADS)

As the name indicates, this system is used to de-pressurize the system. This gets activated in the cases where HPCI alone is not capable of maintaining the requisite coolant level in the core while the reactor pressure is high. To ensure that the coolant level in maintained in the core, supply of coolant from sources other than HPCI must be facilitated. This is done by ADS which reduces pressure by releasing steam to a suppression pool containing water through pipes that connect the steam region with the suppression pool, where the steam is condensed. This brings the reactor pressure down to ~32 atm at which another ECCS subsystem, Low Pressure Coolant Injection (LPCI) system gets activated to supply coolant.

Low Pressure Coolant Injection (LPCI)

This subsystem injects large volumes of water when the reactor pressure falls below 32 atm, through the use of four diesel-driven pumps. This system is capable of flooding the core with water in a short period of time.

This subsystem has several variants depending upon the model of BWR. For example, Advanced Boiling Water Reactor (ABWR) has Low Pressure Core Flooder (LPCF) which functions on similar lines to that of LPCI.

Moderator and Moderator System

We have seen the use of various materials as moderators in different nuclear reactors. One may also recall a reactor classification based on the moderator. That leads us to two important questions: (i) what type of materials is suitable as moderator? (ii) What is the method of comparing the performance of different materials used as moderators?

To answer the first question, we shall recollect different type of nuclear reactions.

During elastic scattering, the speed and direction of a neutron are changed due to elastic collisions with a nucleus. The neutron looses a part of its kinetic energy and hence slows down. These are the thermal neutrons that help the nuclear chain reaction to sustain. This process is called moderation.

The nucleus with which the neutrons undergo elastic collision must be conducive for thermalization. In other words, the probability of elastic reaction between neutron and the nucleus must be high. In terms of cross sections, the nuclear cross section of a moderator's nucleus for elastic scattering (σ_{el}) must be high.

Using the analogy of collisions of billiard balls for elastic scattering, it is possible to relate the mass of target or moderator nucleus (M), energy of incident neutron (E_i) and the energy of scattered neutron (E_s) using the laws of conservation of mass and energy, as follows:

$$E_s = \left(\frac{M - m_n}{M + m_n}\right)^2 E_i$$

If the atomic mass of the target or moderator nucleus is approximated to its mass number (A) and mass of a neutron as 1 amu, then:

$$E_s = \left(\frac{A-1}{A+1}\right)^2 E_i$$

Note: The simultaneous solution of momentum and energy conservation equations for elastic collision between two bodies (as in the case of collision of a neutron with a nucleus) gives two solutions:

Solution – I: Initial and final energies of neutron are same. In this case, there is no change in the direction as neutron continues to travel in forward direction. Hence E_s=E_i.

Solution – II: The neutron is scattered at 180° i.e. the direction of neutron is completely reversed. The energies of incident and scattered neutrons are related by above equation.

In reality, the direction of scattering may range from 0 to 180 °. Hence, the average energy of scattered neutron may be taken as the average of energies with scattering angle 0 and 180°.

Therefore,

$$E_s = \frac{E_i + \left(\frac{A-1}{A+1}\right)^2 E_i}{2} = E_i \frac{1 + \left(\frac{A-1}{A+1}\right)^2}{2}$$

$$\frac{E_s}{E_i} = \frac{1 + \left(\frac{A-1}{A+1}\right)^2}{2} = \frac{A^2 + 1}{A^2 + 2A + 1}$$

The above equation can be graphically represented as shown in figure.

From figure it is clear that lower the value of 'A', lower is the E_s/E_i ratio. Hence light nuclei are more effective in moderating compared to that of heavy nuclei. This is the rationale for using water (hydrogen as scattering nucleus), heavy water (deuterium as scattering nucleus) and graphite (carbon as scattering nucleus).

Above equations give the ratio of kinetic energies of scattered and incident neutron after one collision. However to thermalize the neutrons by reducing its energy from few MeV to 0.0253 eV (8-orders decrease in kinetic energy) the neutron must undergo several successive elastic collisions with the moderator.

The mean logarithmic energy decrement (ξ) defined as $\ln(E_i/E_s)$ is related to the mass number of moderator as:

$$\xi = 1 - \frac{(A-1)^2}{2A}\ln\left(\frac{A+1}{A-1}\right)$$

Higher the value of ξ, more effective the moderation is due to the lower energy of scattered neutron compared to the incident neutron. Equation (5) can be represented diagrammatically in Figure. It is clear from figure that ξ is high for nuclei of lower atomic number.

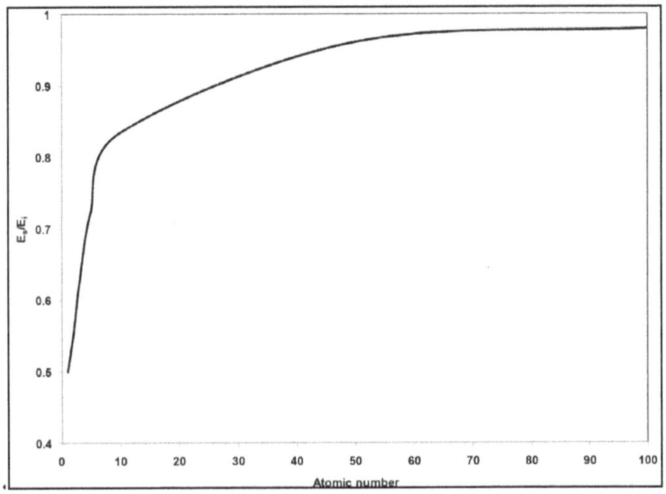

Ratio of energies of scattered neutrons to incident neutrons.

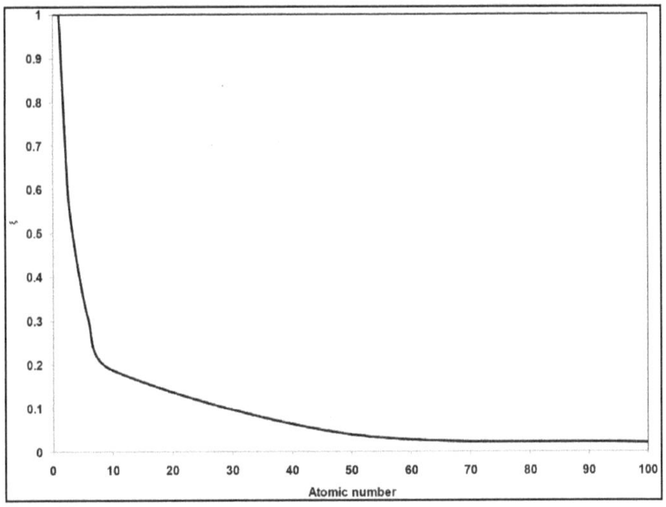

Mean logarithmic energy decrement as a function of atomic number.

If E_{high} and $E_{thermal}$ are the kinetic energies of high energy neutron and thermal neutron respectively, then:

$$\frac{E_{high}}{E_{thermal}} = \frac{E_{high}}{E_1}\cdot\frac{E_1}{E_2}\cdot\frac{E_2}{E_3}\cdots\cdots\frac{E_{n-1}}{E_{thermal}}$$

$$\frac{E_{high}}{E_{thermal}} = e^{\xi}.e^{\xi}.e^{\xi}...e^{\xi} = e^{n\xi}$$

The following example shows the calculation of nuclear of collisions for a typical case.

Example 1: Determine the number of collisions required for thermalization for the following case.

Data: Energy of incident neutron = E_{high} = 1 MeV

Energy of thermal neutron = $E_{thermal}$ = 0.0253 MeV

Moderator: Heavy water,

$\xi = 0.509$

Solution:

Using Eq. (7):

$$\frac{1e6}{0.0253} = e^{0.509n}$$

Solving above Eq., we get n =34

For light water (ξ = 0.920) and graphite (ξ = 0.158) as moderators, the numbers of collisions required for reduction of neutron energy from 1 MeV to 0.0253 eV are 19 and 111 respectively (calculated using earlier equation).

Moderating ratio

It may be recalled that the nuclear cross sections represent the probability of various nuclear reactions. When neutron interacts with a nucleus, one of the following can happen: elastic scattering, inelastic scattering, neutron capture and fission.

As we have seen earlier, the slowing down of neutrons is due to the elastic scattering of neutron by the nucleus of moderator. If the capture cross section of moderator (σ_γ) is high, most of the neutrons will be absorbed by it, leading to lower moderation or lower availability of thermal neutrons.

Hence a higher ratio of scattering to capture cross sections ($\sigma_{el} / \sigma_\gamma$) is desirable for effective moderation. Combining this ratio with mean logarithmic energy decrement (ξ), 'moderating ratio' can be calculated which can be used as a criterion for comparison of different moderators.

$$Moderating\ ratio = \frac{\zeta \sigma_{el}}{\sigma_\gamma}$$

Moderating ratio will be high if either of (ξ) or ($\sigma_{el} / \sigma_\gamma$) is higher. The following table gives the values of ξ, and $\sigma_{el} / \sigma_\gamma$ of some common moderators.

Table: Moderating ratio for common moderators. Note that the cross sections are corresponding to the neutron energy of 0.0253 eV (thermal neutron).

Moderator	ξ	σ_{el} (b)	σ_{γ} (b)	$\sigma_{el}/\sigma_{\gamma}$	Moderating ratio	n
Light water	0.920	25.47	0.33	77.17	71	19
Heavy water	0.509	5.57	0.0005	11139.49	5670	34
Graphite	0.128	5.25	0.0035	1500	192	111

Though light water has the highest ξ and σ_{el} among the three moderators shown, its moderating ratio is low due to its relatively higher capture cross section. Heavy water has the highest moderating ratio owing to its lowest cross section for neutron capture. Graphite, despite possessing the lowest ξ and σ_{el}, fairs better than light water due to its lower σ compared to that of light water.

The relative advantages and disadvantages of heavy water and light water, the common moderators in Indian reactors are as follows:

Table: Relative advantages and disadvantages of light and heavy water as moderator.

Moderator	Advantages	Disadvantages
Heavy water	High moderating ratio and hence can be used with natural uranium fuel.	More expensive; Requires a core of larger diameter to facilitate 34 collisions between a neutron and the moderator.
Light water	Less expensive; Requires a core of relatively smaller diameter as only 19 collisions are required for moderation.	Can be used only with enriched uranium fuel.

Moderator System in Pressurized Heavy Water Reactor

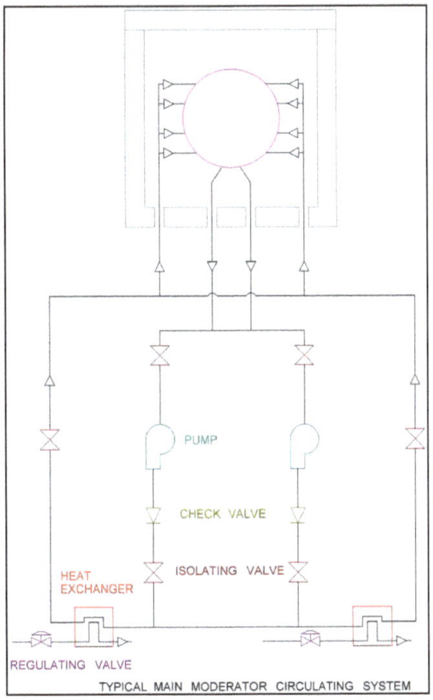

Moderator system of PHWR (Redrawn from Ref.)

Coolant and moderator are not separated in light water reactors. However, despite using heavy water as both moderator and coolant, they are housed in separate compartments in a pressurized heavy water reactor. Heavy water as coolant flows inside the pressure tubes, while the calandria is immersed in the moderator. The moderator is not directly heated by heat transfer from fuel. Instead, moderator is heated due to the energy lost by neutrons during their slow down. The energy of prompt gamma radiation is also deposited in the moderator. It may be recalled that about 5 % of 200 MeV of recoverable energy from fission is released in the moderator. Hence the moderator needs to be cooled to maintain its temperature, to prevent vaporization. Typically, the moderator is maintained between 60 °C and 80°C. The moderator system is designed to maintain the temperature of moderator within this range. A schematic diagram of moderator system with pumps, valves and heat exchangers is shown in Figure.

Calandria has outlets at the bottom for the removal of moderator. The inlets for moderator are provided on the sides of calandria as shown in Figure. Moderator pumps are used to pump the moderator out from calandria and pass them through heat exchangers. Moderator is cooled in these heat exchangers using light water supplied for this purpose. The flow rate of light water is controlled depending upon the temperature of moderator.

Two heat exchangers are shown in Figure. Each heat exchanger is provided with an isolation valve that can be used to disconnect that particular heat exchanger from the system, for maintenance. In such cases, one heat exchanger can be used to maintain the temperature of moderator. The presence of check valves in the line ensure that there is no return flow, in case of failure of moderator pump.

Chapter 7

Nuclear Safety: Concepts and Techniques

Nuclear safety is the achievement of proper operating conditions and prevention or mitigation of accident consequences, resulting in protection of workers, the public and the environment from undue radiation hazards. This chapter provides a detailed analysis into the critical topic of nuclear safety, its concepts and underlying principles and related techniques for preventive, protective and fail-safe measures.

Nuclear Safety and Security

Nuclear safety is defined by the International Atomic Energy Agency (IAEA) as "The achievement of proper operating conditions, prevention of accidents or mitigation of accident consequences, resulting in protection of workers, the public and the environment from undue radiation hazards". The IAEA defines nuclear security as "The prevention and detection of and response to, theft, sabotage, unauthorized access, illegal transfer or other malicious acts involving nuclear material, other radioactive substances or their associated facilities".

A clean-up crew working to remove radioactive contamination after the Three Mile Island accident.

This covers nuclear power plants and all other nuclear facilities, the transportation of nuclear materials, and the use and storage of nuclear materials for medical, power, industry, and military uses.

The nuclear power industry has improved the safety and performance of reactors, and has proposed new and safer reactor designs. However, a perfect safety cannot be guaranteed. Potential sources of problems include human errors and external events that have a greater impact than anticipated: The designers of reactors at Fukushima in Japan did not anticipate that a tsunami generated by an earthquake would disable the backup systems that were supposed to stabilize the reactor after the earthquake. According to UBS AG, the Fukushima I nuclear accidents have cast doubt on whether even an advanced economy like Japan can master nuclear safety. Catastrophic scenarios involving terrorist attacks, insider sabotage, and cyberattacks are also conceivable.

In his book, *Normal accidents*, Charles Perrow says that multiple and unexpected failures are built into society's complex and tightly-coupled nuclear reactor systems. Such accidents are unavoidable and cannot be designed around. To date, there have been three serious accidents (core damage) in the world since 1970, involving five reactors (one at Three Mile Island in 1979; one at Chernobyl in 1986; and three at Fukushima-Daiichi in 2011), corresponding to the beginning of the operation of generation II reactors.

Nuclear weapon safety, as well as the safety of military research involving nuclear materials, is generally handled by agencies different from those that oversee civilian safety, for various reasons, including secrecy. There are ongoing concerns about terrorist groups acquiring nuclear bomb-making material.

Overview of Nuclear Processes and Safety Issues

As of 2011, nuclear safety considerations occur in a number of situations, including:

- Nuclear fission power used in nuclear power stations, and nuclear submarines and ships.
- Nuclear weapons.
- Fissionable fuels such as uranium and plutonium and their extraction, storage and use.
- Radioactive materials used for medical, diagnostc, batteries for some space projects, and research purposes.
- Nuclear waste, the radioactive waste residue of nuclear materials.
- Nuclear fusion power, a technology under long-term development.
- Unplanned entry of nuclear materials into the biosphere and food chain (living plants, animals and humans) if breathed or ingested.

With the exception of thermonuclear weapons and experimental fusion research, all safety issues specific to nuclear power stems from the need to limit the biological uptake of committed dose (ingestion or inhalation of radioactive materials), and external radiation dose due to radioactive contamination.

Nuclear safety therefore covers at minimum:

- Extraction, transportation, storage, processing, and disposal of fissionable materials.

- Safety of nuclear power generators.
- Control and safe management of nuclear weapons, nuclear material capable of use as a weapon, and other radioactive materials.
- Safe handling, accountability and use in industrial, medical and research contexts
- Disposal of nuclear waste.
- Limitations on exposure to radiation.

Responsible Agencies

International

Internationally the International Atomic Energy Agency "works with its Member States and multiple partners worldwide to promote safe, secure and peaceful nuclear technologies." Some scientists say that the 2011 Japanese nuclear accidents have revealed that the nuclear industry lacks sufficient oversight, leading to renewed calls to redefine the mandate of the IAEA so that it can better police nuclear power plants worldwide.

IAEA headquarters in Vienna, Austria.

The IAEA Convention on Nuclear Safety was adopted in Vienna on 17 June 1994 and entered into force on 24 October 1996. The objectives of the Convention are to achieve and maintain a high level of nuclear safety worldwide, to establish and maintain effective defences in nuclear installations against potential radiological hazards, and to prevent accidents having radiological consequences.

The International Atomic Energy Agency was created in 1957 to encourage peaceful development of nuclear technology while providing international safeguards against nuclear proliferation.

The Convention was drawn up in the aftermath of the Three Mile Island and Chernobyl accidents at a series of expert level meetings from 1992 to 1994, and was the result of considerable work by States, including their national regulatory and nuclear safety authorities, and the International Atomic Energy Agency, which serves as the Secretariat for the Convention.

The obligations of the Contracting Parties are based to a large extent on the application of the safety principles for nuclear installations contained in the IAEA document Safety Fundamentals 'The Safety of Nuclear Installations' (IAEA Safety Series No. 110 published 1993). These obligations cover the legislative and regulatory framework, the regulatory body, and technical safety obligations related to, for instance, siting, design, construction, operation, the availability of adequate financial and human resources, the assessment and verification of safety, quality assurance and emergency preparedness.

The convention was amended in 2015 by the Vienna Declaration on Nuclear Safety This resulted in the following principles:

1. New nuclear power plants are to be designed, sited, and constructed, consistent with the objective of preventing accidents in the commissioning and operation and, should an accident occur, mitigating possible releases of radionuclides causing long-term off site contamination and avoiding early radioactive releases or radioactive releases large enough to require long-term protective measures and actions.

2. Comprehensive and systematic safety assessments are to be carried out periodically and regularly for existing installations throughout their lifetime in order to identify safety improvements that are oriented to meet the above objective. Reasonably practicable or achievable safety improvements are to be implemented in a timely manner.

3. National requirements and regulations for addressing this objective throughout the lifetime of nuclear power plants are to take into account the relevant IAEA Safety Standards and, as appropriate, other good practices as identified inter alia in the Review Meetings of the CNS.

There are several problems with the IAEA, says Najmedin Meshkati of University of Southern California, writing in 2011:

"It recommends safety standards, but member states are not required to comply; it promotes nuclear energy, but it also monitors nuclear use; it is the sole global organization overseeing the nuclear energy industry, yet it is also weighed down by checking compliance with the Nuclear Non-Proliferation Treaty (NPT)".

National

Many nations utilizing nuclear power have specialist institutions overseeing and regulating nuclear safety. Civilian nuclear safety in the U.S. is regulated by the Nuclear Regulatory Commission (NRC). However, critics of the nuclear industry complain that the regulatory bodies are too intertwined with the inustries themselves to be effective. The book *The Doomsday Machine* for example, offers a series of examples of national regulators, as they put it 'not regulating, just waving' (a pun on *waiving*) to argue that, in Japan, for example, "regulators and the regulated have long been friends, working together to offset the doubts of a public brought up on the horror of the nuclear bombs". Other examples offered include:

- In the United States, a dangerous custom whereby only supporters of the nuclear industry are allowed to supervise it and lobbyists have been allowed to have an effective veto over regulators.

- In China, where Kang Rixin, former general manager of the state-owned China National Nuclear Corporation, was sentenced to life in jail in 2010 for accepting bribes (and other abuses), a verdict raising questions about the quality of his work on the safety and trustworthiness of China's nuclear reactors.
- In India, where the nuclear regulator reports to the national Atomic Energy Commission, which champions the building of nuclear power plants there and the chairman of the Atomic Energy Regulatory Board, S. S. Bajaj, was previously a senior executive at the Nuclear Power Corporation of India, the company he is now helping to regulate.
- In Japan, where the regulator reports to the Ministry of Economy, Trade and Industry, which overtly seeks to promote the nuclear industry and ministry posts and top jobs in the nuclear business are passed among the same small circle of experts.

The book argues that nuclear safety is compromised by the suspicion that, as Eisaku Sato, formerly a governor of Fukushima province (with its infamous nuclear reactor complex), has put it of the regulators: "They're all birds of a feather".

The safety of nuclear plants and materials controlled by the U.S. government for research, weapons production, and those powering naval vessels is not governed by the NRC. In the UK nuclear safety is regulated by the Office for Nuclear Regulation (ONR) and the Defence Nuclear Safety Regulator (DNSR). The Australian Radiation Protection and Nuclear Safety Agency (ARPANSA) is the Federal Government body that monitors and identifies solar radiation and nuclear radiation risks in Australia. It is the main body dealing with ionizing and non-ionizing radiation and publishes material regarding radiation protection.

Other agencies include:

- Autorité de sûreté nucléaire.
- Canadian Nuclear Safety Commission.
- Radiological Protection Institute of Ireland.
- Federal Atomic Energy Agency in Russia.
- Kernfysische dienst, (NL).
- Pakistan Nuclear Regulatory Authority.
- Bundesamt für Strahlenschutz, (DE).
- Atomic Energy Regulatory Board (India).

Nuclear Power Plant Safety and Security

Complexity

Nuclear power plants are some of the most sophisticated and complex energy systems ever designed. Any complex system, no matter how well it is designed and engineered, cannot be deemed failure-proof. Veteran journalist and author Stephanie Cooke has argued:

The reactors themselves were enormously complex machines with an incalculable number of things that could go wrong. When that happened at Three Mile Island in 1979, another fault line in the nuclear world was exposed. One malfunction led to another, and then to a series of others, until the core of the reactor itself began to melt, and even the world's most highly trained nuclear engineers did not know how to respond. The accident revealed serious deficiencies in a system that was meant to protect public health and safety.

The 1979 Three Mile Island accident inspired Perrow's book *Normal Accidents,* where a nuclear accident occurs, resulting from an unanticipated interaction of multiple failures in a complex system. TMI was an example of a normal accident because it was "unexpected, incomprehensible, uncontrollable and unavoidable".

Perrow concluded that the failure at Three Mile Island was a consequence of the system's immense complexity. Such modern high-risk systems, he realized, were prone to failures however well they were managed. It was inevitable that they would eventually suffer what he termed a 'normal accident'. Therefore, he suggested, we might do better to contemplate a radical redesign, or if that was not possible, to abandon such technology entirely.

A fundamental issue contributing to a nuclear power system's complexity is its extremely long lifetime. The timeframe from the start of construction of a commercial nuclear power station through the safe disposal of its last radioactive waste, may be 100 to 150 years.

Failure Modes of Nuclear Power Plants

There are concerns that a combination of human and mechanical error at a nuclear facility could result in significant harm to people and the environment:

Operating nuclear reactors contain large amounts of radioactive fission products which, if dispersed, can pose a direct radiation hazard, contaminate soil and vegetation, and be ingested by humans and animals. Human exposure at high enough levels can cause both short-term illness and death and longer-term death by cancer and other diseases.

It is impossible for a commercial nuclear reactor to explode like a nuclear bomb since the fuel is never sufficiently enriched for this to occur.

Nuclear reactors can fail in a variety of ways. Should the instability of the nuclear material generate unexpected behavior, it may result in an uncontrolled power excursion. Normally, the cooling system in a reactor is designed to be able to handle the excess heat this causes; however, should the reactor also experience a loss-of-coolant accident, then the fuel may melt or cause the vessel in which it is contained to overheat and melt. This event is called a nuclear meltdown.

After shutting down, for some time the reactor still needs external energy to power its cooling systems. Normally this energy is provided by the power grid to which that plant is connected, or by emergency diesel generators. Failure to provide power for the cooling systems, as happened in Fukushima I, can cause serious accidents.

Nuclear safety rules in the United States "do not adequately weigh the risk of a single event that would knock out electricity from the grid and from emergency generators, as a quake and tsunami recently did in Japan", Nuclear Regulatory Commission officials said in June 2011.

As a safeguard against mechanical failure, many nuclear plants are designed to shut down automatically after two days of continuous and unattended operation.

Vulnerability of Nuclear Plants to Attack

Nuclear reactors become preferred targets during military conflict and, over the past three decades, have been repeatedly attacked during military air strikes, occupations, invasions and campaigns:

- In September 1980, Iran bombed the Al Tuwaitha nuclear complex in Iraq in Operation Scorch Sword.
- In June 1981, an Israeli air strike completely destroyed Iraq's Osirak nuclear research facility in Operation Opera.
- Between 1984 and 1987, Iraq bombed Iran's Bushehr nuclear plant six times.
- On 8 January 1982, Umkhonto we Sizwe, the armed wing of the ANC, attacked South Africa's Koeberg nuclear power plant while it was still under construction.
- In 1991, the U.S. bombed three nuclear reactors and an enrichment pilot facility in Iraq.
- In 1991, Iraq launched Scud missiles at Israel's Dimona nuclear power plant.
- In September 2007, Israel bombed a Syrian reactor under construction.

In the U.S., plants are surrounded by a double row of tall fences which are electronically monitored. The plant grounds are patrolled by a sizeable force of armed guards. The NRC's "Design Basis Threat" criterion for plants is a secret, and so what size of attacking force the plants are able to protect against is unknown. However, to scram (make an emergency shutdown) a plant takes fewer than 5 seconds while unimpeded restart takes hours, severely hampering a terrorist force in a goal to release radioactivity.

Attack from the air is an issue that has been highlighted since the September 11 attacks in the U.S. However, it was in 1972 when three hijackers took control of a domestic passenger flight along the east coast of the U.S. and threatened to crash the plane into a U.S. nuclear weapons plant in Oak Ridge, Tennessee. The plane got as close as 8,000 feet above the site before the hijackers' demands were met.

The most important barrier against the release of radioactivity in the event of an aircraft strike on a nuclear power plant is the containment building and its missile shield. Current NRC Chairman Dale Klein has said "Nuclear power plants are inherently robust structures that our studies show provide adequate protection in a hypothetical attack by an airplane. The NRC has also taken actions that require nuclear power plant operators to be able to manage large fires or explosions—no matter what has caused them."

In addition, supporters point to large studies carried out by the U.S. Electric Power Research Institute that tested the robustness of both reactor and waste fuel storage and found that they should be able to sustain a terrorist attack comparable to the September 11 terrorist attacks in the U.S. Spent fuel is usually housed inside the plant's "protected zone" or a spent nuclear fuel shipping cask; stealing it for use in a "dirty bomb" would be extremely difficult. Exposure to the intense radiation would almost certainly quickly incapacitate or kill anyone who attempts to do so.

Threat of Terrorist Attacks

Nuclear power plants are considered to be targets for terrorist attacks, these findings will be discussed on 11 September not only since the attacks. Even during the construction of the first nuclear power plants has been advised by security bodies on this issue. Even concrete threats of attack against nuclear power plants by terrorists or criminals are documented from several states. While older nuclear power plants were built without special protection against air accidents in Germany, the later nuclear power plants built with a massive concrete buildings are partially protected against air accidents. They are designed against the impact of combat aircraft at a speed of about 800 km / h. It was assumed as a basis of assessment of the impact of an aircraft of type Phantom II with a mass of 20 tonnes and speed of 215 m/ s.

The dangers arising from a terrorist caused large aircraft crash on a nuclear power plant is currently being discussed. Such a terrorist attack could have catastrophic consequences. For example, the German government has confirmed that the nuclear power plant Biblis A not against the crash had secured a military aircraft. Following the terrorist attacks in Brussels in 2016 several nuclear power plants have been partially evacuated. At the same time it became known that the terrorists had spied on the nuclear power plants. Several employees access privileges has been withdrawn.

Moreover, even "nuclear terrorism", for instance with a so-called "Dirty bomb" pose a considerable potential hazard. For their production would come any radioactive waste or enriched for nuclear power plants uranium in question.

Plant Location

In many countries, plants are often located on the coast, in order to provide a ready source of cooling water for the essential service water system. As a consequence the design needs to take the risk of flooding and tsunamis into account. The World Energy Council (WEC) argues disaster risks are changing and increasing the likelihood of disasters such as earthquakes, cyclones, hurricanes, typhoons, flooding. High temperatures, low precipitation levels and severe droughts may lead to fresh water shortages. Failure to calculate the risk of flooding correctly lead to a Level 2 event on the International Nuclear Event Scale during the 1999 Blayais Nuclear Power Plant flood, while flooding caused by the 2011 Tōhoku earthquake and tsunami lead to the Fukushima I nuclear accidents.

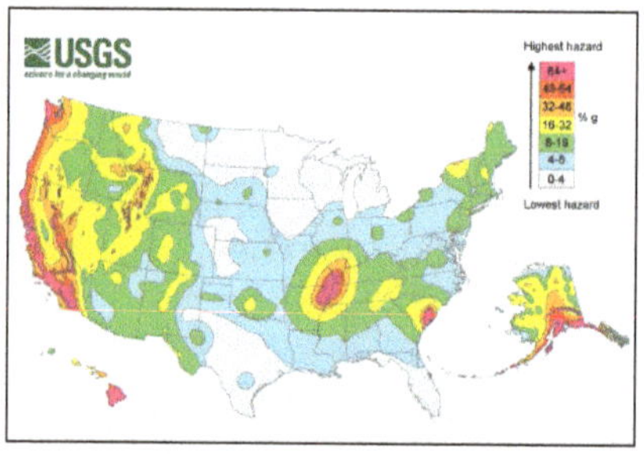

Earthquake map.

Fort Calhoun Nuclear Generating Station surrounded by the 2011 Missouri River Floods on June 16, 2011.

The design of plants located in seismically active zones also requires the risk of earthquakes and tsunamis to be taken into account. Japan, India, China and the USA are among the countries to have plants in earthquake-prone regions. Damage caused to Japan's Kashiwazaki-Kariwa Nuclear Power Plant during the 2007 Chūetsu offshore earthquake underlined concerns expressed by experts in Japan prior to the Fukushima accidents, who have warned of a *genpatsu-shinsai* (domino-effect nuclear power plant earthquake disaster).

Angra Nuclear Power Plant in Rio de Janeiro state, Brazil.

Multiple Reactors

The Fukushima nuclear disaster illustrated the dangers of building multiple nuclear reactor units close to one another. Because of the closeness of the reactors, Plant Director Masao Yoshida "was put in the position of trying to cope simultaneously with core meltdowns at three reactors and exposed fuel pools at three units".

Nuclear Safety Systems

The three primary objectives of nuclear safety systems as defined by the Nuclear Regulatory Commission are to shut down the reactor, maintain it in a shutdown condition, and prevent the release of radioactive material during events and accidents. These objectives are accomplished using a variety of equipment, which is part of different systems, of which each performs specific functions.

Routine Emissions of Radioactive Materials

During everyday routine operations, emissions of radioactive materials from nuclear plants are released to the outside of the plants although they are quite slight amounts. The daily emissions go into the air, water and soil.

NRC says, "nuclear power plants sometimes release radioactive gases and liquids into the environment under controlled, monitored conditions to ensure that they pose no danger to the public or the environment", and "routine emissions during normal operation of a nuclear power plant are never lethal".

According to the United Nations (UNSCEAR), regular nuclear power plant operation including the nuclear fuel cycle amounts to 0.0002 millisieverts (mSv) annually in average public radiation exposure; the legacy of the Chernobyl disaster is 0.002 mSv/a as a global average as of a 2008 report; and natural radiation exposure averages 2.4 mSv annually although frequently varying depending on an individual's location from 1 to 13 mSv.

Japanese Public Perception of Nuclear Power Safety

In March 2012, Prime Minister Yoshihiko Noda said that the Japanese government shared the blame for the Fukushima disaster, saying that officials had been blinded by an image of the country's technological infallibility and were "all too steeped in a safety myth."

Japan has been accused by authors such as journalist Yoichi Funabashi of having an "aversion to facing the potential threat of nuclear emergencies." According to him, a national program to develop robots for use in nuclear emergencies was terminated in midstream because it "smacked too much of underlying danger." Though Japan is a major power in robotics, it had none to send in to Fukushima during the disaster. He mentions that Japan's Nuclear Safety Commission stipulated in its safety guidelines for light-water nuclear facilities that "the potential for extended loss of power need not be considered." However, this kind of extended loss of power to the cooling pumps caused the Fukushima meltdown. In other countries such as the UK, nuclear plants have not been claimed to be absolutely safe. It is instead claimed that a major accident has a likelihood of occurrence lower than (for example) 0.0001/year.

Incidents such as the Fukushima Daiichi nuclear disaster could have been avoided with stricter regulations over nuclear power. In 2002, TEPCO, the company that operated the Fukushima plant, admitted to falsifying reports on over 200 occasions between 1997 and 2002. TEPCO faced no fines for this. Instead, they fired four of their top executives. Three of these four later went on to take jobs at companies that do business with TEPCO.

Hazards of Nuclear Material

There is currently a total of 47,000 tonnes of high-level nuclear waste stored in the USA. Nuclear waste is approximately 94% Uranium, 1.3% Plutonium, 0.14% other Actinides, and 5.2% fission products. About 1.0% of this waste consists of long-lived isotopes ^{79}Se, ^{93}Zr, ^{99}Te, ^{107}Pd, ^{126}Sn, ^{129}I and ^{135}Cs. Shorter lived isotopes including ^{89}Sr, ^{90}Sr, ^{106}Ru, ^{125}Sn, ^{134}Cs, ^{137}Cs, and ^{147}Pm constitute 0.9% at one year, decreasing to 0.1% at 100 years. The remaining 3.3-4.1% consists of non-radioactive isotopes. There are technical challenges, as it is preferable to lock away the long-lived fission products, but the challenge should not be exaggerated. One tonne of waste, as described above, has measurable radioactivity of approximately 600 TBq equal to the natural radioactivity in one km^3 of the Earth's crust, which if buried, would add only 25 parts per trillion to the total radioactivity.

Spent nuclear fuel stored underwater and uncapped at the Hanford site in Washington, USA.

The difference between short-lived high-level nuclear waste and long-lived low-level waste can be illustrated by the following example. As stated above, one mole of both ^{131}I and ^{129}I release $3x10^{23}$ decays in a period equal to one half-life. ^{131}I decays with the release of 970 keV whilst ^{129}I decays with the release of 194 keV of energy. 131gm of ^{131}I would therefore release 45 Gigajoules over eight days beginning at an initial rate of 600 EBq releasing 90 Kilowatts with the last radioactive decay occurring inside two years. In contrast, 129gm of ^{129}I would therefore release 9 Gigajoules over 15.7 million years beginning at an initial rate of 850 MBq releasing 25 microwatts with the radioactivity decreasing by less than 1% in 100,000 years.

One tonne of nuclear waste also reduces CO_2 emission by 25 million tonnes.

Radionuclides such as ^{129}I or ^{131}I, may be highly radioactive, or very long-lived, but they cannot be both. One mole of ^{129}I (129 grams) undergoes the same number of decays ($3x10^{23}$) in 15.7 million years, as does one mole of ^{131}I (131 grams) in 8 days. ^{131}I is therefore highly radioactive, but disappears very quickly, whilst ^{129}I releases a very low level of radiation for a very long time. Two long-lived fission products, Technetium-99 (half-life 220,000 years) and Iodine-129 (half-life 15.7 million years), are of somewhat greater concern because of a greater chance of entering the biosphere. The transuranic elements in spent fuel are Neptunium-237 (half-life two million years) and Plutonium-239 (half-life 24,000 years). will also remain in the environment for long periods of time. A more complete solution to both the problem of both Actinides and to the need for low-carbon energy may be the integral fast reactor. One tonne of nuclear waste after a complete burn in an IFR reactor will have prevented 500 million tonnes of CO_2 from entering the atmosphere. Otherwise, waste storage usually necessitates treatment, followed by a long-term management strategy involving permanent storage, disposal or transformation of the waste into a non-toxic form.

Governments around the world are considering a range of waste management and disposal options, usually involving deep-geologic placement, although there has been limited progress toward implementing long-term waste management solutions. This is partly because the timeframes in question when dealing with radioactive waste range from 10,000 to millions of years, according to studies based on the effect of estimated radiation doses.

Since the fraction of a radioisotope's atoms decaying per unit of time is inversely proportional to its half-life, the relative radioactivity of a quantity of buried human radioactive waste would diminish over time compared to natural radioisotopes (such as the decay chain of 120 trillion tons of thorium and 40 trillion tons of uranium which are at relatively trace concentrations of parts per million each over the crust's $3 * 10^{19}$ ton mass). For instance, over a timeframe of thousands of years, after the most active short half-life radioisotopes decayed, burying U.S. nuclear waste would increase the radioactivity in the top 2000 feet of rock and soil in the United States (10 million km^2) by $\approx$ 1 part in 10 million over the cumulative amount of natural radioisotopes in such a volume, although the vicinity of the site would have a far higher concentration of artificial radioisotopes underground than such an average.

Safety Culture and Human Errors

One relatively prevalent notion in discussions of nuclear safety is that of safety culture. The International Nuclear Safety Advisory Group, defines the term as "the personal dedication and accountability of all individuals engaged in any activity which has a bearing on the safety of nuclear power plants". The goal is "to design systems that use human capabilities in appropriate ways, that protect systems from human frailties, and that protect humans from hazards associated with the system".

At the same time, there is some evidence that operational practices are not easy to change. Operators almost never follow instructions and written procedures exactly, and "the violation of rules appears to be quite rational, given the actual workload and timing constraints under which the operators must do their job". Many attempts to improve nuclear safety culture "were compensated by people adapting to the change in an unpredicted way".

According to Areva's Southeast Asia and Oceania director, Selena Ng, Japan's Fukushima nuclear disaster is "a huge wake-up call for a nuclear industry that hasn't always been sufficiently transparent about safety issues". She said "There was a sort of complacency before Fukushima and I don't think we can afford to have that complacency now".

An assessment conducted by the *Commissariat à l'Énergie Atomique* (CEA) in France concluded that no amount of technical innovation can eliminate the risk of human-induced errors associated with the operation of nuclear power plants. Two types of mistakes were deemed most serious: errors committed during field operations, such as maintenance and testing, that can cause an accident; and human errors made during small accidents that cascade to complete failure.

According to Mycle Schneider, reactor safety depends above all on a 'culture of security', including the quality of maintenance and training, the competence of the operator and the workforce, and the rigour of regulatory oversight. So a better-designed, newer reactor is not always a safer one, and older reactors are not necessarily more dangerous than newer ones. The 1979 Three Mile Island accident in the United States occurred in a reactor that had started operation only three months earlier, and the Chernobyl disaster occurred after only two years of operation. A serious loss of coolant occurred at the French Civaux-1 reactor in 1998, less than five months after start-up.

However safe a plant is designed to be, it is operated by humans who are prone to errors. Laurent Stricker, a nuclear engineer and chairman of the World Association of Nuclear Operators says that operators must guard against complacency and avoid overconfidence. Experts say that the "largest single internal factor determining the safety of a plant is the culture of security among regulators, operators and the workforce — and creating such a culture is not easy".

Risks

The routine health risks and greenhouse gas emissions from nuclear fission power are small relative to those associated with coal, but there are several "catastrophic risks":

The extreme danger of the radioactive material in power plants and of nuclear technology in and of itself is so well known that the US government was prompted (at the industry's urging) to enact provisions that protect the nuclear industry from bearing the full burden of such inherently risky nuclear operations. The Price-Anderson Act limits industry's liability in the case of accidents, and the 1982 Nuclear Waste Policy Act charges the federal government with responsibility for permanently storing nuclear waste.

Population density is one critical lens through which other risks have to be assessed, says Laurent Stricker, a nuclear engineer and chairman of the World Association of Nuclear Operators:

The KANUPP plant in Karachi, Pakistan, has the most people — 8.2 million — living within 30 kilometres of a nuclear plant, although it has just one relatively small reactor with an output of 125 megawatts. Next in the league, however, are much larger plants — Taiwan's 1,933-megawatt Kuosheng plant with 5.5 million people within a 30-kilometre radius and the 1,208-megawatt Chin Shan plant with 4.7 million; both zones include the capital city of Taipei.

172,000 people living within a 30 kilometre radius of the Fukushima Daiichi nuclear power plant, have been forced or advised to evacuate the area. More generally, a 2011 analysis by *Nature* and Columbia University, New York, shows that some 21 nuclear plants have populations larger than 1 million within a 30-km radius, and six plants have populations larger than 3 million within that radius.

Black Swan events are highly unlikely occurrences that have big repercussions. Despite planning, nuclear power will always be vulnerable to black swan events:

A rare event – especially one that has never occurred – is difficult to foresee, expensive to plan for and easy to discount with statistics. Just because something is only supposed to happen every 10,000 years does not mean that it will not happen tomorrow. Over the typical 40-year life of a plant, assumptions can also change, as they did on September 11, 2001, in August 2005 when Hurricane Katrina struck, and in March, 2011, after Fukushima.

The list of potential black swan events is "damningly diverse":

Nuclear reactors and their spent-fuel pools could be targets for terrorists piloting hijacked planes. Reactors may be situated downstream from dams that, should they ever burst, could unleash massive floods. Some reactors are located close to earthquake faults or shorelines, a dangerous scenario like that which emerged at Three Mile Island and Fukushima – a catastrophic coolant failure, the overheating and melting of the radioactive fuel rods, and a release of radioactive material.

- International Nuclear Events Scale.
- Comparative Risk Assessment.
- Statistical Risk Assessment.

- Probabilistic risk assessment:
    - *Severe Accident Risks: An Assessment for Five U.S. Nuclear Power Plants* NUREG-1150 1991.
    - *Calculation of Reactor Accident Consequences* CRAC-II 1982.
    - Rasmussen Report: *Reactor Safety Study* WASH-1400 1975.
    - The Brookhaven Report: *Theoretical Possibilities and Consequences of Major Accidents in Large Nuclear Power Plants* WASH-740 1957.

The AP1000 has a maximum core damage frequency of 5.09×10^{-7} per plant per year. The Evolutionary Power Reactor (EPR) has a maximum core damage frequency of 4×10^{-7} per plant per year. General Electric has recalculated maximum core damage frequencies per year per plant for its nuclear power plant designs:

BWR/4 -- 1×10^{-5}

BWR/6 -- 1×10^{-6}

ABWR -- 2×10^{-7}

ESBWR -- 3×10^{-8}

Beyond Design Basis Events

The Fukushima I nuclear accident was caused by a "beyond design basis event," the tsunami and associated earthquakes were more powerful than the plant was designed to accommodate, and the accident is directly due to the tsunami overflowing the too-low seawall. Since then, the possibility of unforeseen beyond design basis events has been a major concern for plant operators.

Transparency and Ethics

According to journalist Stephanie Cooke, it is difficult to know what really goes on inside nuclear power plants because the industry is shrouded in secrecy. Corporations and governments control what information is made available to the public. Cooke says "when information is made available, it is often couched in jargon and incomprehensible prose".

Kennette Benedict has said that nuclear technology and plant operations continue to lack transparency and to be relatively closed to public view:

Despite victories like the creation of the Atomic Energy Commission, and later the Nuclear Regular Commission, the secrecy that began with the Manhattan Project has tended to permeate the civilian nuclear program, as well as the military and defense programs.

In 1986, Soviet officials held off reporting the Chernobyl disaster for several days. The operators of the Fukushima plant, Tokyo Electric Power Co, were also criticised for not quickly disclosing information on releases of radioactivity from the plant. Russian President Dmitry Medvedev said there must be greater transparency in nuclear emergencies.

Historically many scientists and engineers have made decisions on behalf of potentially affected populations about whether a particular level of risk and uncertainty is acceptable for them. Many nuclear engineers and scientists that have made such decisions, even for good reasons relating to long term energy availability, now consider that doing so without informed consent is wrong, and that nuclear power safety and nuclear technologies should be based fundamentally on morality, rather than purely on technical, economic and business considerations.

Non-Nuclear Futures: The Case for an Ethical Energy Strategy is a 1975 book by Amory B. Lovins and John H. Price. The main theme of the book is that the most important parts of the nuclear power debate are not technical disputes but relate to personal values, and are the legitimate province of every citizen, whether technically trained or not.

Nuclear and Radiation Accidents

The nuclear industry has an excellent safety record and the deaths per megawatt hour are the lowest of all the major energy sources. According to Zia Mian and Alexander Glaser, the "past six decades have shown that nuclear technology does not tolerate error". Nuclear power is perhaps the primary example of what are called 'high-risk technologies' with 'catastrophic potential', because "no matter how effective conventional safety devices are, there is a form of accident that is inevitable, and such accidents are a 'normal' consequence of the system." In short, there is no escape from system failures.

Whatever position one takes in the nuclear power debate, the possibility of catastrophic accidents and consequent economic costs must be considered when nuclear policy and regulations are being framed.

Accident Liability Protection

Kristin Shrader-Frechette has said "if reactors were safe, nuclear industries would not demand government-guaranteed, accident-liability protection, as a condition for their generating electricity". No private insurance company or even consortium of insurance companies "would shoulder the fearsome liabilities arising from severe nuclear accidents".

Hanford Site

The Hanford site represents two-thirds of America's high-level radioactive waste by volume. Nuclear reactors line the riverbank at the Hanford Site along the Columbia River in January 1960.

The Hanford Site is a mostly decommissioned nuclear production complex on the Columbia River in the U.S. state of Washington, operated by the United States federal government. Plutonium manufactured at the site was used in the first nuclear bomb, tested at the Trinity site, and in Fat Man, the bomb detonated over Nagasaki, Japan. During the Cold War, the project was expanded to include nine nuclear reactors and five large plutonium processing complexes, which produced plutonium for most of the 60,000 weapons in the U.S. nuclear arsenal. Many of the early safety procedures and waste disposal practices were inadequate, and government documents have since confirmed that Hanford's operations released significant amounts of radioactive materials into the air and the Columbia River, which still threatens the health of residents and ecosystems. The weapons production reactors were decommissioned at the end of the Cold War, but the decades of manufacturing left behind 53 million US gallons (200,000 m³) of high-level radioactive waste, an additional 25 million cubic feet (710,000 m³) of solid radioactive waste, 200 square miles (520 km²) of contaminated groundwater beneath the site and occasional discoveries of undocumented contaminations that slow the pace and raise the cost of cleanup. The Hanford site represents two-thirds of the nation's high-level radioactive waste by volume. Today, Hanford is the most contaminated nuclear site in the United States and is the focus of the nation's largest environmental cleanup.

1986 Chernobyl Disaster

The Chernobyl disaster was a nuclear accident that occurred on 26 April 1986 at the Chernobyl Nuclear Power Plant in Ukraine. An explosion and fire released large quantities of radioactive contamination into the atmosphere, which spread over much of Western USSR and Europe. It is considered the worst nuclear power plant accident in history, and is one of only two classified as a level 7 event on the International Nuclear Event Scale (the other being the Fukushima Daiichi nuclear disaster). The battle to contain the contamination and avert a greater catastrophe ultimately involved over 500,000 workers and cost an estimated 18 billion rubles, crippling the Soviet economy. The accident raised concerns about the safety of the nuclear power industry, slowing its expansion for a number of years.

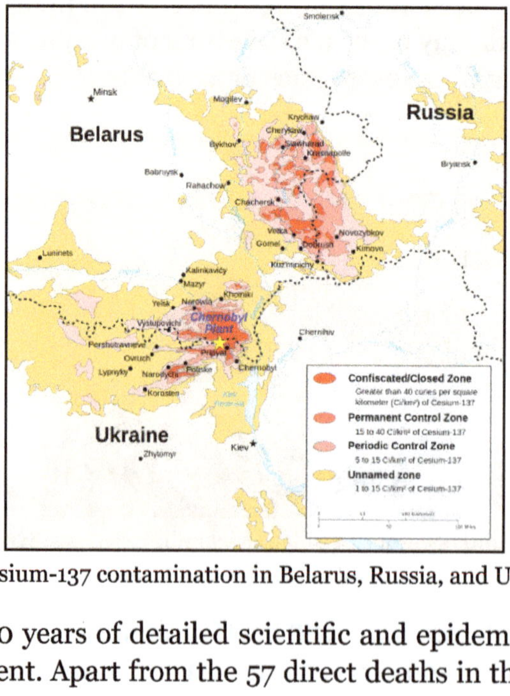

Map showing Caesium-137 contamination in Belarus, Russia, and Ukraine as of 1996.

UNSCEAR has conducted 20 years of detailed scientific and epidemiological research on the effects of the Chernobyl accident. Apart from the 57 direct deaths in the accident itself, UNSCEAR predicted in 2005 that up to 4,000 additional cancer deaths related to the accident would appear "among the 600 000 persons receiving more significant exposures (liquidators working in 1986–87, evacuees, and residents of the most contaminated areas)". Russia, Ukraine, and Belarus have been burdened with the continuing and substantial decontamination and health care costs of the Chernobyl disaster.

Eleven of Russia's reactors are of the RBMK 1000 type, similar to the one at Chernobyl Nuclear Power Plant. Some of these RBMK reactors were originally to be shut down but have instead been given life extensions and uprated in output by about 5%. Critics say that these reactors are of an "inherently unsafe design", which cannot be improved through upgrades and modernization, and some reactor parts are impossible to replace. Russian environmental groups say that the lifetime extensions "violate Russian law, because the projects have not undergone environmental assessments".

2011 Fukushima I Accidents

Despite all assurances, a major nuclear accident on the scale of the 1986 Chernobyl disaster happened again in 2011 in Japan, one of the world's most industrially advanced countries. Nuclear Safety Commission Chairman Haruki Madarame told a parliamentary inquiry in February 2012 that "Japan's atomic safety rules are inferior to global standards and left the country unprepared for the Fukushima nuclear disaster last March". There were flaws in, and lax enforcement of, the safety rules governing Japanese nuclear power companies, and this included insufficient protection against tsunamis.

Fukushima reactor control room.

Following the 2011 Japanese Fukushima nuclear disaster, authorities shut down the nation's 54 nuclear power plants. As of 2013, the Fukushima site remains highly radioactive, with some 160,000 evacuees still living in temporary housing, and some land will be unfarmable for centuries. The difficult cleanup job will take 40 or more years, and cost tens of billions of dollars.

A 2012 report in *The Economist* said: "The reactors at Fukushima were of an old design. The risks they faced had not been well analysed. The operating company was poorly regulated and did not know what was going on. The operators made mistakes. The representatives of the safety inspectorate fled. Some of the equipment failed. The establishment repeatedly played down the risks and suppressed information about the movement of the radioactive plume, so some people were evacuated from more lightly to more heavily contaminated places".

The designers of the Fukushima I Nuclear Power Plant reactors did not anticipate that a tsunami generated by an earthquake would disable the backup systems that were supposed to stabilize the reactor after the earthquake. Nuclear reactors are such "inherently complex, tightly coupled systems that, in rare, emergency situations, cascading interactions will unfold very rapidly in such a way that human operators will be unable to predict and master them".

Lacking electricity to pump water needed to cool the atomic core, engineers vented radioactive steam into the atmosphere to release pressure, leading to a series of explosions that blew out concrete walls around the reactors. Radiation readings spiked around Fukushima as the disaster widened, forcing the evacuation of 200,000 people. There was a rise in radiation levels on the outskirts of Tokyo, with a population of 30 million, 135 miles (210 kilometers) to the south.

Back-up diesel generators that might have averted the disaster were positioned in a basement, where they were quickly overwhelmed by waves. The cascade of events at Fukushima had been predicted in a report published in the U.S. several decades ago:

The 1990 report by the U.S. Nuclear Regulatory Commission, an independent agency responsible for safety at the country's power plants, identified earthquake-induced diesel generator failure and power outage leading to failure of cooling systems as one of the "most likely causes" of nuclear accidents from an external event.

The report was cited in a 2004 statement by Japan's Nuclear and Industrial Safety Agency, but it seems adequate measures to address the risk were not taken by TEPCO. Katsuhiko Ishibashi, a seismology professor at Kobe University, has said that Japan's history of nuclear accidents stems from an overconfidence in plant engineering. In 2006, he resigned from a government panel on nuclear reactor safety, because the review process was rigged and "unscientific".

According to the International Atomic Energy Agency, Japan "underestimated the danger of tsunamis and failed to prepare adequate backup systems at the Fukushima Daiichi nuclear plant". This repeated a widely held criticism in Japan that "collusive ties between regulators and industry led to weak oversight and a failure to ensure adequate safety levels at the plant". The IAEA also said that the Fukushima disaster exposed the lack of adequate backup systems at the plant. Once power was completely lost, critical functions like the cooling system shut down. Three of the reactors "quickly overheated, causing meltdowns that eventually led to explosions, which hurled large amounts of radioactive material into the air".

Louise Fréchette and Trevor Findlay have said that more effort is needed to ensure nuclear safety and improve responses to accidents:

The multiple reactor crises at Japan's Fukushima nuclear power plant reinforce the need for strengthening global instruments to ensure nuclear safety worldwide. The fact that a country that has been operating nuclear power reactors for decades should prove so alarmingly improvisational in its response and so unwilling to reveal the facts even to its own people, much less the International Atomic Energy Agency, is a reminder that nuclear safety is a constant work-in-progress.

David Lochbaum, chief nuclear safety officer with the Union of Concerned Scientists, has repeatedly questioned the safety of the Fukushima I Plant's General Electric Mark 1 reactor design, which is used in almost a quarter of the United States' nuclear fleet.

A report from the Japanese Government to the IAEA says the "nuclear fuel in three reactors probably melted through the inner containment vessels, not just the core". The report says the "inadequate" basic reactor design — the Mark-1 model developed by General Electric — included "the venting system for the containment vessels and the location of spent fuel cooling pools high in the buildings, which resulted in leaks of radioactive water that hampered repair work".

Following the Fukushima emergency, the European Union decided that reactors across all 27 member nations should undergo safety tests.

According to UBS AG, the Fukushima I nuclear accidents are likely to hurt the nuclear power industry's credibility more than the Chernobyl disaster in 1986:

The accident in the former Soviet Union 25 years ago 'affected one reactor in a totalitarian state with no safety culture,' UBS analysts including Per Lekander and Stephen Oldfield wrote in a report today. 'At Fukushima, four reactors have been out of control for weeks -- casting doubt on whether even an advanced economy can master nuclear safety.'

The Fukushima accident exposed some troubling nuclear safety issues:

Despite the resources poured into analyzing crustal movements and having expert committees determine earthquake risk, for instance, researchers never considered the possibility of a magnitude-9 earthquake followed by a massive tsunami. The failure of multiple safety features on nuclear power plants has raised questions about the nation's engineering prowess. Government flip-flopping on acceptable levels of radiation exposure confused the public, and health professionals provided little guidance. Facing a dearth of reliable information on radiation levels, citizens armed themselves with dosimeters, pooled data, and together produced radiological contamination maps far more detailed than anything the government or official scientific sources ever provided.

As of January 2012, questions also linger as to the extent of damage to the Fukushima plant caused by the earthquake even before the tsunami hit. Any evidence of serious quake damage at the plant would "cast new doubt on the safety of other reactors in quake-prone Japan".

Two government advisers have said that "Japan's safety review of nuclear reactors after the Fukushima disaster is based on faulty criteria and many people involved have conflicts of interest". Hiromitsu Ino, Professor Emeritus at the University of Tokyo, says "The whole process being undertaken is exactly the same as that used previous to the Fukushima Dai-Ichi accident, even though the accident showed all these guidelines and categories to be insufficient".

In March 2012, Prime Minister Yoshihiko Noda acknowledged that the Japanese government shared the blame for the Fukushima disaster, saying that officials had been blinded by a false belief in the country's "technological infallibility", and were all too steeped in a "safety myth".

Other Accidents

Serious nuclear and radiation accidents include the Chalk River accidents (1952, 1958 & 2008), Mayak disaster (1957), Windscale fire (1957), SL-1 accident (1961), Soviet submarine K-19 accident (1961), Three Mile Island accident (1979), Church Rock uranium mill spill (1979), Soviet submarine K-431 accident (1985), Goiânia accident (1987), Zaragoza radiotherapy accident (1990), Costa Rica radiotherapy accident (1996), Tokaimura nuclear accident (1999), Sellafield THORP leak (2005), and the Flerus IRE cobalt-60 spill (2006).

Health Impacts

Four hundred and thirty-seven nuclear power stations are presently in operation but, unfortunately, five major nuclear accidents have occurred in the past. These accidents occurred at Kyshtym (1957), Windscale (1957), Three Mile Island (1979), Chernobyl (1986), and Fukushima (2011). A report in *Lancet* says that the effects of these accidents on individuals and societies are diverse and enduring:

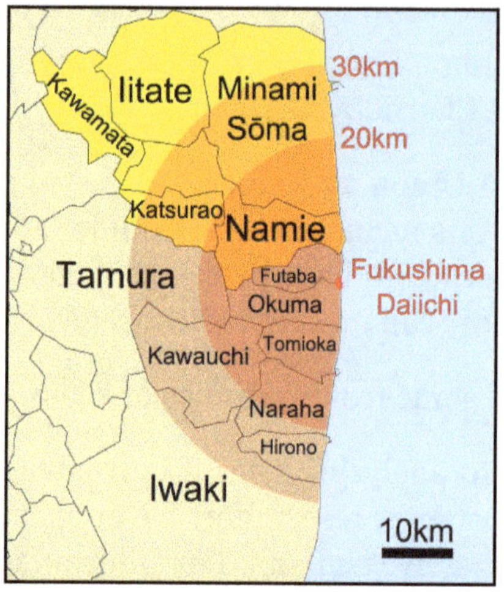

Japan towns, villages, and cities around the Fukushima Daiichi nuclear plant. The 20km and 30km areas had evacuation and sheltering orders, and additional administrative districts that had an evacuation order are highlighted.

> "Accumulated evidence about radiation health effects on atomic bomb survivors and other radiation-exposed people has formed the basis for national and international regulations about radiation protection. However, past experiences suggest that common issues were not necessarily physical health problems directly attributable to radiation exposure, but rather psychological and social effects. Additionally, evacuation and long-term displacement created severe health-care problems for the most vulnerable people, such as hospital inpatients and elderly people."

In spite of accidents like these, studies have shown that nuclear deaths are mostly in uranium mining and that nuclear energy has generated far fewer deaths than the high pollution levels that result from the use of conventional fossil fuels. However, the nuclear power industry relies on uranium mining, which itself is a hazardous industry, with many accidents and fatalities.

Journalist Stephanie Cooke says that it is not useful to make comparisons just in terms of number of deaths, as the way people live afterwards is also relevant, as in the case of the 2011 Japanese nuclear accidents.

"You have people in Japan right now that are facing either not returning to their homes forever, or if they do return to their homes, living in a contaminated area for basically ever... It affects millions of people, it affects our land, it affects our atmosphere ... it's affecting future generations ... I don't think any of these great big massive plants that spew pollution into the air are good. But I don't think it's really helpful to make these comparisons just in terms of number of deaths".

The Fukushima accident forced more than 80,000 residents to evacuate from neighborhoods around the plant.

A survey by the Iitate, Fukushima local government obtained responses from some 1,743 people who have evacuated from the village, which lies within the emergency evacuation zone around the crippled Fukushima Daiichi Plant. It shows that many residents are experiencing growing frustration and instability due to the nuclear crisis and an inability to return to the lives they were living before the disaster. Sixty percent of respondents stated that their health and the health of their families had deteriorated after evacuating, while 39.9 percent reported feeling more irritated compared to before the disaster.

"Summarizing all responses to questions related to evacuees' current family status, one-third of all surveyed families live apart from their children, while 50.1 percent live away from other family members (including elderly parents) with whom they lived before the disaster. The survey also showed that 34.7 percent of the evacuees have suffered salary cuts of 50 percent or more since the outbreak of the nuclear disaster. A total of 36.8 percent reported a lack of sleep, while 17.9 percent reported smoking or drinking more than before they evacuated."

Chemical components of the radioactive waste may lead to cancer. For example, Iodine 131 was released along with the radioactive waste when Chernobyl disaster and Three Mile Island accidents occurred. It was concentrated in leafy vegetation after absorption in the soil. It also stays in animals' milk if the animals eat the vegetation. When Iodine 131 enters the human body, it migrates to the thyroid gland in the neck and can cause thyroid cancer.

Other elements from nuclear waste can lead to cancer as well. For example, Strontium 90 causes breast cancer and leukemia, Plutonium 239 causes liver cancer.

Improvements to Nuclear Fission Technologies

Newer reactor designs intended to provide increased safety have been developed over time. These designs include those that incorporate passive safety and Small Modular Reactors. While these reactor designs "are intended to inspire trust, they may have an unintended effect: creating distrust of older reactors that lack the touted safety features".

The next nuclear plants to be built will likely be Generation III or III+ designs, and a few such are already in operation in Japan. Generation IV reactors would have even greater improvements in safety. These new designs are expected to be passively safe or nearly so, and perhaps even inherently safe (as in the PBMR designs).

Some improvements made (not all in all designs) are having three sets of emergency diesel generators and associated emergency core cooling systems rather than just one pair, having quench tanks (large coolant-filled tanks) above the core that open into it automatically, having a double containment (one containment building inside another), etc.

However, safety risks may be the greatest when nuclear systems are the newest, and operators have less experience with them. Nuclear engineer David Lochbaum explained that almost all serious nuclear accidents occurred with what was at the time the most recent technology. He argues that "the problem with new reactors and accidents is twofold: scenarios arise that are impossible to plan for in simulations; and humans make mistakes". As one director of a U.S. research laboratory put it, "fabrication, construction, operation, and maintenance of new reactors will face a steep learning curve: advanced technologies will have a heightened risk of accidents and mistakes. The technology may be proven, but people are not".

Developing Countries

There are concerns about developing countries "rushing to join the so-called nuclear renaissance without the necessary infrastructure, personnel, regulatory frameworks and safety culture". Some countries with nuclear aspirations, like Nigeria, Kenya, Bangladesh and Venezuela, have no significant industrial experience and will require at least a decade of preparation even before breaking ground at a reactor site.

The speed of the nuclear construction program in China has raised safety concerns. The challenge for the government and nuclear companies is to "keep an eye on a growing army of contractors and subcontractors who may be tempted to cut corners". China is advised to maintain nuclear safeguards in a business culture where quality and safety are sometimes sacrificed in favor of cost-cutting, profits, and corruption. China has asked for international assistance in training more nuclear power plant inspectors.

Nuclear Security and Terrorist Attacks

Nuclear power plants, civilian research reactors, certain naval fuel facilities, uranium enrichment plants, and fuel fabrication plants, are vulnerable to attacks which could lead to widespread radioactive contamination. The attack threat is of several general types: commando-like ground-based attacks on equipment which if disabled could lead to a reactor core meltdown or widespread dispersal of radioactivity; and external attacks such as an aircraft crash into a reactor complex, or cyber attacks.

The United States 9/11 Commission has said that nuclear power plants were potential targets originally considered for the September 11, 2001 attacks. If terrorist groups could sufficiently damage safety systems to cause a core meltdown at a nuclear power plant, and/or sufficiently damage spent fuel pools, such an attack could lead to widespread radioactive contamination. The Federation of American Scientists have said that if nuclear power use is to expand significantly, nuclear facilities will have to be made extremely safe from attacks that could release massive quantities of radioactivity into the community. New reactor designs have features of passive safety, which may help. In the United States, the NRC carries out "Force on Force" (FOF) exercises at all Nuclear Power Plant (NPP) sites at least once every three years.

Nuclear reactors become preferred targets during military conflict and, over the past three decades, have been repeatedly attacked during military air strikes, occupations, invasions and campaigns. Various acts of civil disobedience since 1980 by the peace group Plowshares have shown how nuclear weapons facilities can be penetrated, and the groups actions represent extraordinary breaches of security at nuclear weapons plants in the United States. The National Nuclear Security Administration has acknowledged the seriousness of the 2012 Plowshares action. Non-proliferation policy experts have questioned "the use of private contractors to provide security at facilities that manufacture and store the government's most dangerous military material". Nuclear weapons materials on the black market are a global concern, and there is concern about the possible detonation of a small, crude nuclear weapon by a militant group in a major city, with significant loss of life and property. *Stuxnet* is a computer worm discovered in June 2010 that is believed to have been created by the United States and Israel to attack Iran's nuclear facilities.

Nuclear Fusion Research

Nuclear fusion power is a developing technology still under research. It relies on fusing rather than fissioning (splitting) atomic nuclei, using very different processes compared to current nuclear power plants. Nuclear fusion reactions have the potential to be safer and generate less radioactive waste than fission. These reactions appear potentially viable, though technically quite difficult and have yet to be created on a scale that could be used in a functional power plant. Fusion power has been under theoretical and experimental investigation since the 1950s.

Construction of the International Thermonuclear Experimental Reactor facility began in 2007, but the project has run into many delays and budget overruns. The facility is now not expected to begin operations until the year 2027 – 11 years after initially anticipated. A follow on commercial nuclear fusion power station, DEMO, has been proposed. There is also suggestions for a power plant based upon a different fusion approach, that of a Inertial fusion power plant.

Fusion powered electricity generation was initially believed to be readily achievable, as fission power had been. However, the extreme requirements for continuous reactions and plasma containment led to projections being extended by several decades. In 2010, more than 60 years after the first attempts, commercial power production was still believed to be unlikely before 2050.

More Stringent Safety Standards

Matthew Bunn, the former US Office of Science and Technology Policy adviser, and Heinonen, the former Deputy Director General of the IAEA, have said that there is a need for more stringent nuclear safety standards, and propose six major areas for improvement:

- Operators must plan for events beyond design bases;
- More stringent standards for protecting nuclear facilities against terrorist sabotage;
- A stronger international emergency response;
- International reviews of security and safety;
- Binding international standards on safety and security;
- International co-operation to ensure regulatory effectiveness.

Coastal nuclear sites must also be further protected against rising sea levels, storm surges, flooding, and possible eventual "nuclear site islanding".

Nuclear Decommissioning

Nuclear decommissioning is the process whereby a nuclear power plant site is dismantled to the point that it no longer requires measures for radiation protection. The presence of radioactive material necessitates processes that are occupationally dangerous, hazardous to the natural environment, expensive, and time-intensive.

The reactor pressure vessel being transported away from the site for burial.

Example of decommissioning work underway.

Decommissioning is an administrative and technical process. It includes clean-up of radioactive materials and progressive demolition of the plant. Once a facility is fully decommissioned, no radiologic danger should persist. The costs of decommissioning are spread over the lifetime of a facility and saved in a decommissioning fund. After a facility has been completely decommissioned, it is released from regulatory control and the plant licensee is no longer responsible for its nuclear safety. Decommissioning may proceed all the way to "greenfield" status.

Options

The International Atomic Energy Agency has defined three options for decommissioning:

- *Immediate Dismantling* (Early Site Release/DECON in the US): This option allows for the facility to be removed from regulatory control relatively soon after shutdown or termination of regulated activities. Final dismantling or decontamination activities begin within a few months or years, and depending on the facility, it could take five years or more. Following removal from regulatory control, the site becomes available for re-use.

- *Safe Enclosure* (or Safestor(e) SAFSTOR): This option postpones the final removal of controls for a longer period, usually on the order of 40 to 60 years. The facility is placed into a safe storage configuration until the eventual dismantling and decontamination activities occur.

- *Entombment/ENTOMB*: This option entails placing the facility into a condition that will allow the remaining radioactive material to remain on-site indefinitely. This option usually involves reducing the size of the area where the radioactive material is located and then encasing the facility in a long-lived material such as concrete, theoretically preventing a release of radioactive material.

Experience

A wide range of nuclear facilities have been decommissioned so far. This includes nuclear power plants (NPPs), research reactors, isotope production plants, particle accelerators, and uranium mines. The number of decommissioned power plants is small. Companies specialize in nuclear decommissioning; decommissioning has become a profitable business. More recently, construction and demolition companies in the UK have also begun to develop nuclear decommissioning services. The current estimate by the United Kingdom's Nuclear Decommissioning Authority is that it will cost at least £100 billion to decommission the 19 existing United Kingdom nuclear sites. Due to the radioactivity in the reactor structure, decommissioning takes place in stages. The plans of the Nuclear Decommissioning Authority for decommissioning reactors have an average 50 year time frame. The long time frame makes reliable cost estimates difficult. Cost overruns are common even for quick projects.

North America

The Pickering Nuclear Generating Station, viewed from the west. All eight reactors are visible; two units have been shut down.

Most nuclear plants currently operating in the United States were designed for a life of about 30–40 years and are licensed to operate for 40 years by the US Nuclear Regulatory Commission. The average age of these reactors is 32 years. Many are coming to the end of their licensing period. If their licenses are not renewed, the plants must go through a decontamination and decommissioning process.

Several nuclear reactors dismantled in North America, type, power, and decommissioning cost:

Dismantled nuclear reactors in Canada & USA					
Country:	**Location:**	**Reactor type:**	**Operative life:**	**Decommissioning phase:**	**Dismantling costs:**
Canada (Québec)	Gentilly-1	CANDU-BWR 250 MWe	180 days (between 1966 and 1973)	"Static state" since 1986	stage two: $25 million
Canada (Ontario)	Pickering NGS Units A2 and A3	CANDU-PWR 8 x 542 MWe	30 years (from 1974 to 2004)	Two units currently in "cold standby" Decommissioning to begin in 2020	(calculated: $270–430/ kWe ?)

United States	Fort St. Vrain	HTGR (helium-graphite) 380 MWe	12 years (1977–1989)	Immediate Decon	$195 million
USA	Rancho Seco	Multiunit: PWR 913 MWe	12 years (Closed after a referendum in 1989)	SAFSTOR: 5–10 years completion 2018	? ($200–500/ kWe)
USA	Three Mile Island 2	Multiunit: 913 MWe PWR	INCIDENT: core fusion (in 1979)	Post-Defuelling Phase 2 (1979)	$805 Million (estimated)
USA	Shippingport	(The first BWR) 60 MWe	25 years (closed in 1989)	Decon completed dismantled in 5 years (first small experimental reactor)	$98.4 million
USA	Piqua(Ohio)	OCM (Organically Cooled/ Moderated) reactor 46 MWe	2 years (closed in 1966)	ENTOMB (coolant design inadequate for neutron flux)	?
USA	Trojan	PWR 1,180 MWe	16 years (Closed in 1993 because nearby to seismic fault)	SAFSTOR (cooling tower demolished in 2006)	?
USA	Yankee Rowe	PWR 185 MWe	31 years (1960–1991)	Decon completed - Demolished (greenfield open to visitors)	$608 million with $8 million per year upkeep
USA	Maine Yankee	PWR 860 MWe	24 years (closed in 1996)	Decon completed - Demolished in 2004 (greenfield open to visitors)	$635 million
USA	Connecticut Yankee	PWR 590 MWe	28 years (closed in 1996)	Decon - Demolished in 2007 (greenfield open to visitors)	$820 million
USA	Exelon - Zion 1 & 2	PWR - Westinghouse 2 x 1040 MWe	25 years (1973–1998) (Incident in proceedings, abandoned because of the excessive cost of vaporizers substitution)	SAFSTOR- EnergySolutions (opening of the site to visitors for 2018)	$900–1,100 million (2007 dollars)

USA	Pacific Gas & Electric - Humboldt Bay Nuclear Power Plant - Unit 3	BWR 1 x 63 MWe	13 years (1963–1976) (Shut down due to seismic retrofit)	On July 2, 1976, Humboldt Bay Power Plant (HBPP) Unit 3 was shut down for annual refueling and to conduct seismic modifications. In 1983, updated economic analyses indicated that restarting Unit 3 would probably not be cost-effective, and in June 1983, PG&E announced its intention to decommission the unit. On July 16, 1985, the U.S. Nuclear Regulatory Commission (NRC) issued Amendment No. 19 to the HBPP Unit 3 Operating License to change the status to possess-but-not-operate, and the plant was placed into a SAFSTOR status.	Unknown - Closure date: 12/31/2015

Asia

Reactors Not Located in Japan

Several nuclear reactors dismantled in Asia, type, power and decommissioning cost per kilowatt of electric power (source: World Nuclear Association).

Dismantled reactors in Asia					
Country:	**Location:**	**Reactor type:**	**Operative life:**	**Decommissioning phase:**	**Dismantling cost:**
China	Beijing (CIAE)	HWWR 10 MWe (multipurpose) (Heavy Water Experimental Reactor for the production of plutonium and tritium)	49 years (1958–2007)	SAFSTOR & Decon in 20 years (until 2027)	proposed: $6 Million for dismantling $5 Million for fuel remotion
North Korea	Yongbyon	Magnox-type (reactor for the production of nuclear weapons through PUREX treatment)	20 years (1985–2005) Deactivated after a treaty	SAFSTOR: cooling tower dismantled	?

India	Tarapur-1,2 (Maharashtra)	2x BWR 160 MWe	40 years ? (1969–2009?)	NOT deactivated	?
India	Rawatbhata Atomic Power Station-1,2 (Rajasthan)	1x PHWR 100 MWe 1x PHWR 200 MWe (similar to CANDU)	40 years ? (1970–2011?)	NOT deactivated	?
Iraq	Osiraq/Tammuz-1	BWR 40 MWe Nuclear reactor with weapons-grade plutonium production capability	(Destroyed by Israeli Air Force in 1981)	Not radioactive: never refurbished with uranium	?

Japan

The three damaged reactors at Fukushima Dai-ichi #1,#2,#3 are expected to be decommissioned as well as #4.

Heavily damaged reactors in Japan					
Nuclear Power Plant	**Electric max. output** (MW)	**Type**	**Connection to electric grid**	**Situation**	**Decommissioning costs**
Fukushima Dai-ichi NPP (Unit 1)	439	BWR	November 17, 1970	nuclear meltdown Hydrogen explosion (INES 7)	? Estimated at ¥10 trillion (US$ 100 billion) for decontaminating Fukushima and dismantling all reactors in Japan and considering long time damage to environment and economy, including agriculture, cattle breeding, fishery, water depuration, tourism (without considering further health care spending & reduction of life expectancy).
Fukushima Dai-ichi NPP (Unit 2)	760	BWR	December 24, 1973	partial nuclear meltdown (INES 6) (Risk of going into INES 7)	?
Fukushima Dai-ichi NPP (Unit 3)	760	BWR	October 26, 1974	nuclear meltdown Hydrogen explosion (INES 7)	?

Fukushima Dai-ichi NPP (Unit 4)	760	BWR	February 24, 1978	Reactor defuelled when tsunami hit Damage to spent fuel cooling-pool (INES 4) (Could worsen if pool collapses) (Other specialists disagree about this danger)	?
Fukushima Dai-ichi NPP (Unit 5)	760	BWR	September 22, 1977	SCRAM	?
Fukushima Daiichi NPP (Unit 6)	1067	BWR	May 4, 1979	SCRAM	?
Fukushima Daini NPP (Unit 1)	1067	BWR	July 31, 1981	SCRAM (leakage of coolant)	?
Fukushima Daini NPP (Units 2 - 4)	3 × 1067	BWR	June 23, 1983 December 14, 1984 December 17, 1986	3 x SCRAM	?
Tokai NPP (Reactor 2)	1100 MW	BWR/5	November 28, 1978	SCRAM Shutdown since March 2011 (anti-tsunami barrier stopped the waves) INES 1 (leakage of coolant)	?

Safely decommissioned reactors in Japan					
Nuclear Power Plant	**Electric max. output** (MW)	**Type**	**Connection to electric grid**	**Situation**	**Decommissioning costs**
Tokai NPP (Reactor 1)	160 MWe	Magnox (GCR)	(1966–1998)	Safstore: 10 years then decon until 2018	¥93 billion (€660 million of 2003)

Western Europe and Former Yugoslavia

Several nuclear reactors dismantled in Western Europe, type, power and decommissioning cost per kilowatt of power: European Union Website about Nuclear Decommissioning, World Nuclear Association (reactor building companies), United Kingdom.

Safely decommissioned reactors in Western Europe					
Country:	**Location:**	**Reactor type:**	**Operative Life:**	**Decommissioning phase:**	**Dismantling cost:**
Austria (Nuclear Free Country)	Zwentendorf NPP Google Maps	PWR 723 MWe	Never activated due to referendum in 1978	?	?

Belgium	SCK•CEN - BR3, located at Mol, Belgium	PWR (BR-3)	25 years (1962–1987)	Decon completed (2011) European pilot project (underwater cutting and remote operated tools)	?
France	Brennilis	HWGCR 70 MWe	12 years (1967–1979)	Phase 3	€480 million (20 times the forecast-ed amount)
France	Bugey-1	UNGG Gas cooled, graphite mod-erator	1972–1994	postponed	?
France	Chinon 1,2,3	Gas-graphite	(1973–1990)	postponed	?
France	Chooz-A	PWR 300 MW	24 years (1967–1991)	Fully decommissioned - Greenfield (Nuclear reactor was lo-cated inside a mountain cave)	?
France	Saint-Laurent Nucle-ar Power Plant	Gas-graphite	1969–1992 50 kg of Uranium in one of the reactors at the Saint-Laurent Nuclear Power Plant began to melt, an event classified at 'level 4' on the International Nuclear Event Scale (INES). As of March 2011, this is the most serious civil nuclear power accident in France.	postponed	?
France	Rapsodie at Cadarache	Experimental Fast breeder nuclear reactor (sodium-cooled) 40 MWe	15 years (1967–1983)	1983: Defuelling 1987: Remotion of neu-tron reflectors 1985-1989: Decontam-ination of sodium coolant Accident when cleaning residual sodium in vessel with ethyl carbitol (March 31° 1994)	The removed activity is estimated to around 4800 TBq. 600 TBq (^{60}Co) in 1990 still contained in 1ry vessel The dose burden from 1987 to 1994 was 224 mSv. RAPSODIE reached IAEA level 2 of de-comm in 2005 STAGE 3 is planned in 2020
France	Phénix at Marcoule	Experimental Fast breeder nuclear reactor (sodium-cooled) 233 MWe	36 years (1973–2009)	1) Defuelled	estimated for the future: $4000/kWe

France	Superphénix at Creys-Malville	Fast breeder nuclear reactor (sodium-cooled)	11 years (1985–1996)	1) Defuelled 2) Extraction of Sodium Pipe cutting with a robot	estimated for the future: $4000/kWe
United Kingdom	Berkeley	Magnox (2 x 138 MWe)	27 years (1962–1989)	SAFSTOR: 30 years (internal demolition)	$2600/kWe
United Kingdom	Sellafield-Windscale	Windscale Advanced Gas Reactor WAGR (32 MWe)	18 years (1963–1981)	Remotion of reactor in 2009 pilot project (cutting with remote controlled robots, UV lasers)	More than $2600/kWe (WNI estimates) So far €117 million
United Kingdom	Dorset -Winfrith Operated in Dorset from 1958 to 1990. Had 9 reactors all mostly dismantled				
West Germany	Gundremmingen-A	BWR 250 MWe	11 years	Immediate dismantling pilot project (underwater cutting)	(~ $300–550/kWe)
Italy	Caorso NPP	BWR 840 MWe	3 years (1978 - Closed in 1987 after referendum in 1986)	SAFSTOR: 30 years (internal demolition)	€450 million (dismantling) + €300 million (fuel reprocessing)
Italy	Garigliano NPP (Caserta)	BWR 150 MWe	? (Closed on March 1, 1982)	SAFSTOR: 30 years (internal demolition)	?
Italy	Latina NPP (Foce Verde)	Magnox 210 MWe Gas-graphite	24 years (1962 - Closed in 1986 after referendum)	SAFSTOR: 30 years (internal demolition)	?
Italy	Trino Vercellese NPP	PWR Westinghouse, 270 MWe	? (Closed in 1986 after referendum)	SAFSTOR: 30 years (internal demolition)	?
Netherlands	Dodewaard NPP	BWR Westinghouse 58 MWe	28 years (1969–1997)	Defuelling completed SAFSTOR: 40 years	?
Spain	Vandellós NPP-1	UNGG 480 MWe (gas-graphite)	18 years Incident: fire in a turbo-generator (1989)	SAFSTOR: 30 years (internal demolition)	Phases 1 and 2: €93 million
Sweden	Barsebäck NPP 1 & 2	BWR 2 x 615 MW	Reactor 1: 24 years 1975 - 1999 Reactor 2: 28 years 1977 - 2005	SAFSTOR: demolition will begin in 2020	The Swedish Radiation Safety Authority has assessed that the costs for decommissioning and final disposal for the Swedish nuclear power industry may be underestimated by SKB by at least 11 billion Swedish crowns ($1.63 billion)

Switzerland	DIORIT	MWe Gas-graphite (experimental)	?	SAFSTOR (internal demolition)	?
Switzerland	LUCENS	8,3 MWe CO_2- heavy water (experimental)	(1962–1969) Incident: fire in 1969	Entombment for ? years SAFSTOR & Decon: 24 years (internal demolition)	?
Switzerland	SAPHIR	0,01-0,1 MWe (Light water pool)	39 years (1955–1994) (Experimental demonstrator)	In public display since inauguration	?

- Repository for radioactive waste Morsleben: €2.2 billion

Eastern Europe and Former Soviet Union

Several nuclear reactors dismantled in the nations born from the former Soviet Union: (Belarus, Russia, Ukraine and others) and reactors dismantled in countries formerly belonging to "Warsaw Pact" and/or to "Comecon", type, electric power and decommissioning cost per kilowatt of power: World Nuclear Association, OSTI (Russia & USA).

Decommissioned reactors in Eastern Europe					
Country:	**Location:**	**Reactor typr:**	**Operative life:**	**Decommissioning phase:**	**Dismantling cost:**
Bulgaria	Kozloduy NPP-1,2,3,4	PWR VVER-440 (4 x 408 MWe)	Reactors 1,2 closed in 2003, reactors 3,4 closed in 2006 (Closing forced by European Union)	De-fuelling	?
East Germany	Greifswald NPP-1, 2,3,4,5,6	VVER-440 5 x 408 MWe	Reactors 1-5 closed in 1989/1990, reactor 6: finished but never operated	Immediate dismantling (underwater cutting)	~ \$330/kWe
East Germany	Rheinsberg NPP-1	VVER-210 70–80 MWe	24 years (1966–1990)	In dismantling since 1996 Safstor (underwater cutting)	~ \$330/kWe
East Germany	Stendal NPP-1,2,3,4	VVER-1000 (4 x 1000 MWe)	Never activated (1st reactor 85% completed)	Not radioactive (Cooling towers demolished with explosives)	? (Structure in exhibition inside an industrial park)
Russia	Mayak (Chelyabinsk-65)	PUREX plant for uranium enrichment	Several severe incidents (1946–1956)	?	?
Russia	Seversk (Tomsk-7)	Three plutonium reactors Plant for uranium enrichment	Two fast-breeder reactors closed (of three), after disarmaments agreements with USA in 2003.	?	?

Slovakia	Jaslovske Bohunice NPP-1,2 (180 km east from Vienna)	VVER 440/230 2 X 440 MWe	(1978–2006) (1980–2008)	?	?
Ukraine	Chernobyl NPP-4 (110 km from Kiev)	RBMK-1000 1000 MWe	hydrogen explosion, then graphite fire (1986) (INES 7)	ENTOMBMENT (armed concrete "sarcophagus")	Past: ? Future: riding sarcophagus in steel

Legal Aspects

The decommission of a nuclear reactor can only take place after the appropriate licence has been granted pursuant to the relevant legislation. As part of the licensing procedure, various documents, reports and expert opinions have to be written and delivered to the competent authority, e.g. safety report, technical documents and an environmental impact study (EIS).

In the European Union these documents are the basis for the environmental impact assessment (EIA) according to Council Directive 85/337/EEC. A precondition for granting such a licence is an opinion by the European Commission according to Article 37 of the Euratom Treaty. Article 37 obliges every Member State of the European Union to communicate certain data relating to the release of radioactive substances to the Commission. This information must reveal whether and if so what radiological impacts decommissioning – planned disposal and accidental release – will have on the environment, i.e. water, soil or airspace, of the EU Member States. On the basis of these general data, the Commission must be in a position to assess the exposure of reference groups of the population in the nearest neighbouring states.

Cost

In USA many utility estimates now average $325 million per reactor all-up (1998 $).

In France, decommissioning of Brennilis Nuclear Power Plant, a fairly small 70 MW power plant, already cost €480 million (20x the estimate costs) and is still pending after 20 years. Despite the huge investments in securing the dismantlement, radioactive elements such as plutonium, caesium-137 and cobalt-60 leaked out into the surrounding lake.

In the UK, decommissioning of the Windscale Advanced Gas Cooled Reactor (WAGR), a 32 MW prototype power plant, cost €117 million.

In Germany, decommissioning of Niederaichbach nuclear power plant, a 100 MW power plant, amounted to more than €143 million.

New methods for decommissioning have been developed in order to minimize the usual high decommissioning costs. One of these methods is in situ decommissioning (ISD), meaning that the reactor is entombed instead of dismantled. This method was implemented at the U.S. Department of Energy Savannah River Site in South Carolina for the closures of the P and R Reactors. With this tactic, the cost of decommissioning both reactors was $73 million. In comparison, the decommissioning of each reactor using traditional methods would have been an estimated $250 million. This results in a 71% decrease in cost by using ISD.

Decommissioning Funds

In Europe there is considerable concern over the funds necessary to finance final decommissioning. In many countries either the funds do not appear sufficient to cover decommissioning and in other countries decommissioning funds are used for other activities, putting decommissioning at risk, and distorting competition with parties who do not have such funds available.

In 2016 the European Commission assessed that European Union's nuclear decommissioning liabilities were seriously underfunded by about 118 billion euros, with only 150 billion euros of earmarked assets to cover 268 billion euros of expected decommissioning costs covering both dismantling of nuclear plants and storage of radioactive parts and waste. France had the largest shortfall with only 23 billion euros of earmarked assets to cover 74 billion euros of expected costs.

Similar concerns exist in the United States, where the U.S. Nuclear Regulatory Commission has located apparent decommissioning funding assurance shortfalls and requested 18 power plants to address that issue. The decommissioning cost of Small Modular Reactors is expected to be twice as much respect to Large Reactors.

International Collaboration

Organizations that promote the international sharing of information, knowledge, and experiences related to nuclear decommissioning include the International Atomic Energy Agency, the Organization for Economic Co-operation and Development's Nuclear Energy Agency and the European Atomic Energy Community. In addition, an online system called the Deactivation and Decommissioning Knowledge Management Information Tool was developed under the United States Department of Energy and made available to the international community to support the exchange of ideas and information. The goals of international collaboration in nuclear decommissioning are to reduce decommissioning costs and improve worker safety.

Ships, Mobile Reactors, Military Reactors

Many warships and a few civil ships have used nuclear reactors for propulsion. Former Soviet and American warships have been taken out of service and their power plants removed or scuttled. Dismantling of Russian submarines and ships and American submarines and ships is ongoing. Marine power plants are generally smaller than land-based electrical generating stations.

Reactor Protection System

A reactor protection system (RPS) is a set of nuclear safety components in a nuclear power plant designed to safely shut down the reactor and prevent the release of radioactive materials. The system can "trip" automatically (initiating a scram), or it can be tripped by the operators. Trips occurs when the parameters meet or exceed the limit setpoint. A trip of the RPS results in full insertion (by gravity in pressurized water reactors or high-speed injection in boiling water reactors) of all control rods and shutdown of the reactor.

Pressurized Water Reactors

Some of the measured parameters for US pressurized water plants would include:

- "High power", auctioneered between high nuclear power and high differential temperature (delta T) between the inlet and outlet of the reactor vessel (a measure of the thermal power for a given RCS flowrate).
- "High startup rate" (active below 10-4 percent power) at low power levels.
- "High pressurizer pressure".
- "Low reactor coolant flow".
- "Thermal margin / low pressure" (reactor power versus RCS pressure).
- "High containment pressure".
- "Low steam generator level".
- "Low steam generator pressure".
- "Loss of load" (main turbine trip).

Each parameter is measured by independent channels such that actuation of any two channels would result in an automatic SCRAM or reactor shutdown. The system also allows manual actuation by the operator.

Passive Nuclear Safety

Passive nuclear safety is a safety feature of a nuclear reactor that does not require operator actions or electronic feedback in order to shut down safely in the event of a particular type of emergency (usually overheating resulting from a loss of coolant or loss of coolant flow). Such reactors tend to rely more on the engineering of components such that their predicted behaviour according to known laws of physics would slow, rather than accelerate, the nuclear reaction in such circumstances. This is in contrast to older-yet-common reactor designs, where the natural tendency for the reaction was to accelerate rapidly from increased temperatures, such that either electronic feedback or operator triggered intervention was necessary to prevent damage to the reactor. The term "walk away" safety is also used to describe this feature.

Terminology

Terming a reactor 'passively safe' is more a description of the strategy used in maintaining a degree of safety, than it is a description of the level of safety. Whether a reactor employing passive safety systems is to be considered safe or dangerous will depend on the criteria used to evaluate the safety level. This said, modern reactor designs have focused on increasing the amount of passive safety, and thus most passively safe designs incorporate both active and passive safety systems, making them substantially safer than older installations. They can be said to be "relatively safe" compared to previous designs.

Reactor vendors like to call their new generation reactors 'passively safe' but this term is sometimes confused with 'inherently safe' in the public perception. It is very important to understand that there are no 'passively safe' reactors or 'passively safe' systems, only 'passively safe' *components* of safety systems exist. Safety systems are used to maintain control of the plant if it goes outside normal conditions in case of anticipated operational occurrences or accidents, while the control systems are used to operate the plant under normal conditions. Sometimes a system combines both features. Passive safety refers to safety system components, whereas inherent safety refers to control system process regardless of the presence or absence of safety specific subsystems.

As an example of a safety system with 'passively safe' components, let us consider the containment of a nuclear reactor. 'Passively safe' components are the concrete walls and the steel liner, but in order to fulfil its mission active systems have to operate, e.g. valves to ensure the closure of the piping leading outside the containment, feedback of reactor status to external instrumentation and control (I&C) both of which may require external power to function.

The International Atomic Energy Agency (IAEA) classifies the degree of "passive safety" of components from category A to D depending on what the system does not make use of:

1. No moving working fluid.
2. No moving mechanical part.
3. No signal inputs of 'intelligence'.
4. No external power input or forces.

In category A (1+2+3+4) is the fuel cladding, the protective and nonreactive outer layer of the fuel pellet, which uses none of the above features: It is always closed and keeps the fuel and the fission products inside and is not open before arriving at the reprocessing plant. In category B (2+3+4) is the surge line, which connects the hot leg with the pressurizer and helps to control the pressure in the primary loop of a PWR and uses a moving working fluid when fulfilling its mission. In category C (3+4) is the accumulator, which does not need signal input of 'intelligence' or external power. Once the pressure in the primary circuit drops below the set point of the spring-loaded accumulator valves, the valves open and water is injected into the primary circuit by compressed nitrogen. In category D (4 only) is the SCRAM which utilizes moving working fluids, moving mechanical parts and signal inputs of 'intelligence' but not external power or forces: the control rods drop driven by gravity once they have been released from their magnetic clamp. But nuclear safety engineering is never that simple: Once released the rod may not fulfil its mission: It may get stuck due to earthquake conditions or due to deformed core structures. This shows that though it is a passively safe system and has been properly actuated, it may not fulfil its mission. Nuclear engineers have taken this into consideration: Typically only a part of the rods dropped are necessary to shut down the reactor. Samples of safety systems with passive safety components can be found in almost all nuclear power stations: the containment, hydro-accumulators in PWRs or pressure suppression systems in BWRs.

In most texts on 'passively safe' components in next generation reactors, the key issue is that no pumps are needed to fulfil the mission of a safety system and that all active components (generally I&C and valves) of the systems work with the electric power from batteries.

IAEA explicitly uses the following caveat:

... passivity is not synonymous with reliability or availability, even less with assured adequacy of the safety feature, though several factors potentially adverse to performance can be more easily counteracted through passive design (public perception). On the other hand active designs employing variable controls permit much more precise accomplishment of safety functions; this may be particularly desirable under accident management conditions.

Nuclear reactor response properties such as Temperature coefficient of reactivity and Void coefficient of reactivity usually refer to the thermodynamic and phase-change response of the neutron moderator heat transfer *process* respectively. Reactors whose heat transfer process has the operational property of a negative void coefficient of reactivity are said to possess an *inherent safety* process feature. An operational failure mode could potentially alter the process to render such a reactor unsafe.

Reactors could be fitted with a hydraulic safety system component that increases the inflow pressure of coolant (esp. water) in response to increased outflow pressure of the moderator and coolant without control system intervention. Such reactors would be described as fitted with such a *passive safety* component that could - if so designed - render in a reactor a negative void coefficient of reactivity, regardless of the operational property of the reactor in which it is fitted. The feature would only work if it responded faster than an emerging (steam) void and the reactor components could sustain the increased coolant pressure. A reactor fitted with both safety features - if designed to constructively interact - is an example of a safety interlock. Rarer operational failure modes could render both such safety features useless and detract from the overall relative safety of the reactor.

Examples of Passive Safety in Operation

Traditional reactor safety systems are *active* in the sense that they involve electrical or mechanical operation on command systems (e.g., high-pressure water pumps). But some engineered reactor systems operate entirely passively, e.g., using pressure relief valves to manage overpressure. Parallel redundant systems are still required. Combined *inherent* and *passive* safety depends only on physical phenomena such as pressure differentials, convection, gravity or the *natural* response of materials to high temperatures to slow or shut down the reaction, not on the functioning of engineered components such as high-pressure water pumps.

Current pressurized water reactors and boiling water reactors are systems that have been designed with one kind of passive safety feature. In the event of an excessive-power condition, as the water in the nuclear reactor core boils, pockets of steam are formed. These steam voids moderate fewer neutrons, causing the power level inside the reactor to lower. The BORAX experiments and the SL-1 meltdown accident proved this principle.

A reactor design whose *inherently* safe process directly provides a *passive* safety component during a specific failure condition in *all* operational modes is typically described as relatively fail-safe to that failure condition. However most current water-cooled and -moderated reactors, when scrammed, can not remove residual production and decay heat without either process heat transfer or the active cooling system. In other words, whilst the inherently safe heat transfer process provides a passive safety component preventing excessive heat while the reactor is operating, the same inherently safe heat transfer process *does not* provide a passive safety component is shut down (SCRAMed). The Three Mile Island accident exposed this design deficiency: the reactor and steam generator were shut down but with loss of coolant it still suffered a partial meltdown.

Third generation designs improve on early designs by incorporating passive or inherent safety features which require *no* active controls or (human) operational intervention to avoid accidents in the event of malfunction, and may rely on pressure differentials, gravity, natural convection, or the natural response of materials to high temperatures.

In some designs the core of a fast breeder reactor is immersed into a pool of liquid metal. If the reactor overheats, thermal expansion of the metallic fuel and cladding causes more neutrons to escape the core, and the nuclear chain reaction can no longer be sustained. The large mass of liquid metal also acts as a heatsink capable of absorbing the decay heat from the core, even if the normal cooling systems would fail.

The pebble bed reactor is an example of a reactor exhibiting an inherently safe process that is also capable of providing a passive safety component for all operational modes. As the temperature of the *fuel* rises, Doppler broadening increases the probability that neutrons are captured by U-238 atoms. This reduces the chance that the neutrons are captured by U-235 atoms and initiate fission, thus reducing the reactor's power output and placing an inherent upper limit on the temperature of the fuel. The geometry and design of the fuel pebbles provides an important passive safety component.

Single fluid fluoride molten salt reactors feature fissile, fertile and actinide radioisotopes in molecular bonds with the fluoride coolant. The molecular bonds provide a passive safety feature in that a loss-of-coolant event corresponds with a loss-of-fuel event. The molten fluoride fuel can not itself reach criticality but only reaches criticality by the addition of a neutron reflector such as pyrolytic graphite. The higher density of the fuel along with additional lower density FLiBe fluoride coolant without fuel provides a flotation layer passive safety component in which lower density graphite that breaks off control rods or an immersion matrix during mechanical failure does not induce criticality. Gravity driven drainage of reactor liquids provides a passive safety component.

Low power pool-type reactors such as the SLOWPOKE and TRIGA have been licensed for *unattended* operation in research environments because as the temperature of the low-enriched (19.75% U-235) uranium alloy hydride fuel rises, the molecular bound hydrogen in the fuel cause the heat to be transferred to the fission neutrons as they are ejected. This Doppler shifting or spectrum hardening dissipates heat from the fuel more rapidly throughout the pool the higher the fuel temperature increases ensuring rapid cooling of fuel whilst maintaining a much lower water temperature than the fuel. Prompt, self-dispersing, high efficiency hydrogen-neutron heat transfer rather than inefficient radionuclide-water heat transfer ensures the fuel cannot melt through accident alone. In uranium-zirconium alloy hydride variants, the fuel itself is also chemically corrosion resistant ensuring a sustainable safety performance of the fuel molecules throughout their lifetime. A large expanse of water and the concrete surround provided by the pool for high energy neutrons to penetrate ensures the process has a high degree of intrinsic safety. The core is visible through the pool and verification measurements can be made directly on the core fuel elements facilitating total surveillance and providing nuclear non-proliferation safety. Both the fuel molecules themselves and the open expanse of the pool are passive safety components. Quality implementations of these designs are arguably the safest nuclear reactors.

Examples of Reactors Using Passive Safety Features

Three Mile Island Unit 2 was unable to contain about 480 PBq of radioactive noble gases from release into the environment and around 120 kL of radioactive contaminated cooling water from release beyond the containment into a neighbouring building. The pilot-operated relief valve at TMI-2 was designed to shut automatically after relieving excessive pressure inside the reactor into a quench tank. However the valve mechanically failed causing the PORV quench tank to fill, and the relief diaphragm to eventually rupture into the containment building. The containment building sump pumps automatically pumped the contaminated water outside the containment building. Both a working PORV with quench tank and separately the containment building with sump provided two layers of passive safety. An unreliable PORV negated its designed passive safety. The plant design featured only a single open/close indicator based on the status of its solenoid actuator, instead of a separate indicator of the PORV's actual position. This rendered the mechanical reliability of the PORV indeterminate directly, and therefore its passive safety status indeterminate. The automatic sump pumps and/or insufficient containment sump capacity negated the containment building designed passive safety.

The notorious RBMK graphite moderated, water-cooled reactors of Chernobyl Power Plant disaster were designed with a positive void coefficient with boron control rods on electromagnetic grapples for reaction speed control. To the degree that the control systems were reliable, this *design* did have a corresponding degree of *active* inherent safety. The reactor was unsafe at low power levels because erroneous control rod movement would have a counter-intuitively magnified effect. Chernobyl Reactor 4 was built instead with manual crane driven boron control rods that were tipped with the moderator substance, graphite, a neutron reflector. It was designed with an Emergency Core Cooling System (ECCS) that depended on either grid power or the backup Diesel generator to be operating. The ECCS safety component was decidedly not passive. The design featured a partial containment consisting of a concrete slab above and below the reactor - with pipes and rods penetrating, an inert gas filled metal vessel to keep oxygen away from the water-cooled hot graphite, a fire-proof roof, and the pipes below the vessel sealed in secondary water filled boxes. The roof, metal vessel, concrete slabs and water boxes are examples of passive safety components. The roof in the Chernobyl Power Plant complex was made of bitumen - against design - rendering it ignitable. Unlike the Three Mile Island accident, neither the concrete slabs nor the metal vessel could contain a steam, graphite and oxygen driven hydrogen explosion. The water boxes could not sustain high pressure failure of the pipes. The passive safety components as designed were inadequate to fulfill the safety requirements of the system.

The General Electric Company ESBWR (Economic Simplified Boiling Water Reactor, a BWR) is a design reported to use passive safety components. In the event of coolant loss, no operator action is required for three days.

The Westinghouse AP1000 ("AP" standing for "Advanced Passive") uses passive safety components. In the event of an accident, no operator action is required for 72 hours. Recent version of the Russian VVER have added a passive heat removal system to the existing active systems, utilising a cooling system and water tanks built on top of the containment dome.

The integral fast reactor was a fast breeder reactor run by the Argonne National Laboratory. It was a sodium cooled reactor capable of withstanding a loss of (coolant) flow without SCRAM and loss of heatsink without SCRAM. This was demonstrated throughout a series of safety tests in which the reactor successfully shut down without operator intervention. The project was canceled due to proliferation concerns before it could be copied elsewhere.

The Molten-Salt Reactor Experiment[13] (MSRE) was a molten salt reactor run by the Oak Ridge National Laboratory. It was nuclear graphite moderated and the coolant salt used was FLiBe, which also carried the uranium-233 fluoride fuel dissolved in it. The MSRE had a negative temperature coefficient of reactivity: as the FLiBe temperature increased, it expanded, along with the uranium ions it carried; this decreased density resulted in a reduction of fissile material in the core, which decreased the rate of fission. With less heat input, the net result was that the reactor would cool. Extending from the bottom of the reactor core was a pipe that lead to passively cooled drain tanks. The pipe had a "freeze valve" along its length, in which the molten salt was actively cooled to a solid plug by a fan blowing air over the pipe. If the reactor vessel developed excessive heat or lost electric power to the air cooling, the plug would melt; the FLiBe would be pulled out of the reactor core by gravity into dump tanks, and criticality would cease as the salt lost contact with the graphite moderator.

The General Atomics HTGR design features a fully passive and inherently safe decay heat removal system, termed the Reactor Cavity Cooling System (RCCS). In this design, an array of steel ducts line the concrete containment (and hence surround the reactor pressure vessel) which provide a flow path for air driven natural circulation from chimneys positioned above grade. Derivatives of this RCCS concept (with either air or water as the working fluid) has also been featured in other gas-cooled reactor designs, including the Japanese HTTR, the Chinese HTR-10, the South African PBMR, and the Russian GT-MHR. While none of these designs have been commercialized for power generation research in these areas is active, specifically in support of the Generation IV initiative and NGNP programs, with experimental facilities at Argonne National Laboratory (home to the Natural convection Shutdown heat removal Test Facility, a 1/2 scale air-cooled RCCS)[14] and the University of Wisconsin (home to separate 1/4 scale air and water-cooled RCCS),.[15][16]

References

- House of Commons Committee of Public Accounts (4 February 2013). "Nuclear Decommissioning Authority: Managing risk at Sellafield" (PDF). London: The Stationery Office Limited. Retrieved 20, January 2020
- "What happens to radiation produced by a plant?". NRC: Frequently Asked Questions (FAQ) About Radiation Protection. Nuclear Regulatory Commission. Retrieved 07, April 2020
- Hiroko Tabuchi (March 3, 2012). "Japanese Prime Minister Says Government Shares Blame for Nuclear Disaster". The New York Times. Retrieved 16, July 2020
- Ojovan, M. I.; Lee, W.E. (2005). An Introduction to Nuclear Waste Immobilisation. Amsterdam: Elsevier Science Publishers. p. 315. ISBN 0-08-044462-8
- "What you can do to protect yourself: Be Informed". Nuclear Power Plants | RadTown USA | US EPA. United States Environmental Protection Agency. Retrieved 05, May 2020
- "Hanford Site: Hanford Overview". United States Department of Energy. Archived from the original on 2012-06-05. Retrieved 19, February 2020
- "Is radiation exposure from a nuclear power plant always fatal?". NRC: Frequently Asked Questions (FAQ) About Radiation Protection. Nuclear Regulatory Commission. Retrieved 17, June 2020

- "How old are U.S. nuclear power plants and when was the last one built? - FAQ - U.S. Energy Information Administration (EIA)". Eia.gov. Retrieved 26, July 2020
- "Radiation Dose". Factsheets & FAQs: Radiation in Everyday Life. International Atomic Energy Agency (IAEA). Retrieved 22, August 2020

Permissions

All chapters in this book are published with permission under the Creative Commons Attribution Share Alike License or equivalent. Every chapter published in this book has been scrutinized by our experts. Their significance has been extensively debated. The topics covered herein carry significant information for a comprehensive understanding. They may even be implemented as practical applications or may be referred to as a beginning point for further studies.

We would like to thank the editorial team for lending their expertise to make the book truly unique. They have played a crucial role in the development of this book. Without their invaluable contributions this book wouldn't have been possible. They have made vital efforts to compile up to date information on the varied aspects of this subject to make this book a valuable addition to the collection of many professionals and students.

This book was conceptualized with the vision of imparting up-to-date and integrated information in this field. To ensure the same, a matchless editorial board was set up. Every individual on the board went through rigorous rounds of assessment to prove their worth. After which they invested a large part of their time researching and compiling the most relevant data for our readers.

The editorial board has been involved in producing this book since its inception. They have spent rigorous hours researching and exploring the diverse topics which have resulted in the successful publishing of this book. They have passed on their knowledge of decades through this book. To expedite this challenging task, the publisher supported the team at every step. A small team of assistant editors was also appointed to further simplify the editing procedure and attain best results for the readers.

Apart from the editorial board, the designing team has also invested a significant amount of their time in understanding the subject and creating the most relevant covers. They scrutinized every image to scout for the most suitable representation of the subject and create an appropriate cover for the book.

The publishing team has been an ardent support to the editorial, designing and production team. Their endless efforts to recruit the best for this project, has resulted in the accomplishment of this book. They are a veteran in the field of academics and their pool of knowledge is as vast as their experience in printing. Their expertise and guidance has proved useful at every step. Their uncompromising quality standards have made this book an exceptional effort. Their encouragement from time to time has been an inspiration for everyone.

The publisher and the editorial board hope that this book will prove to be a valuable piece of knowledge for students, practitioners and scholars across the globe.

Index

www.ingramcontent.com/pod-product-compliance
Lightning Source LLC
LaVergne TN
LVHW061510220926
848485LV00017B/624